# BARRON'S
## NEW JERSEY

# ASK 3
# MATH
# TEST

**SECOND EDITION**

**Thomas P. Walsh, M.A., M.Ed., Ed.D.**
**Dan M. Nale, MBA, M.Ed.**

**Acknowledgments**

I would like to thank my wife, Elaine, for her understanding, great criticisms, and support. I would like to thank my sons, Matthew and Dan, who helped me with ideas. I would also like to thank everyone at Kean University who have encouraged me through this project. —Tom Walsh

I would like to thank my wife, Teresa, for her understanding, support, and enthusiasm. Thank you to my parents, Joel and Michelle, for providing a foundation for my successful endeavors. I would also like to thank the Somers Point school district for providing me with the opportunity to become a third-grade teacher. —Dan M. Nale

New Jersey Core Curriculum Content Standards (NJCCCS), set out on pages 4 to 14, are © copyright New Jersey Department of Education.
Format of Sample Tests in Chapter 9 © copyright NJDOE.

*All inquiries should be addressed to:*
Barron's Educational Series, Inc.
250 Wireless Boulevard
Hauppauge, New York 11788
**www.barronseduc.com**

ISBN-978-1-4380-0191-3
ISSN 2168-832x

Date of Manufacture: December 2012
Manufactured by: B11R11, Robbinsville, NJ

Printed in the United States of America
9 8 7 6 5 4 3 2 1

**10%**
**POST-CONSUMER WASTE**
Paper contains a minimum of 10% post-consumer waste (PCW). Paper used in this book was derived from certified, sustainable forestlands.

# CONTENTS

## 5 MEASUREMENT AND DATA / 73

## 6 GEOMETRY / 123

## 7 PROBLEM SOLVING AND OTHER MATHEMATICAL PROCESSES / 157

## 8 ANSWERS TO PRACTICE PROBLEMS / 183

**SAMPLE TESTS AND ANSWERS / 231**

**IMPORTANT NOTE**

The material in this book reflects The New Jersey Core State Standards as of press time. Because core curriculum standards continue to develop, you should always consult with a teacher with questions. The material in this book will provide a strong framework for all students preparing for the New Jersey ASK 3 Math Exam.

# Chapter 1

# INTRODUCTION TO THE NEW JERSEY ASK 3 MATH TEST FOR PARENTS AND TEACHERS

## SPECIFICATIONS OF THE TEST

**Important!**

This chapter contains technical information geared toward parents and educators. Beginning with chapter 2, students can review the material and complete sample math problems themselves.

When first established, the tests known as the New Jersey Assessment of Skills and Knowledge (NJ ASK) was based on the New Jersey Core Curriculum Content Standards (NJ CCCS). Since the introduction of the NJ ASK, a new set of standards have been adopted by the State of New Jersey. These are called the Common Core State Standards (CCSS). The CCSS were developed to better align standards among the United States with regard to English and mathematics. The effort for reform was led by the Council of Chief State School Officers (CCSSO) and the National Governors Association (NGA). The English and mathematics standards are the beginning; standards in other subject areas (science, history/social studies) will be forthcoming. From the Mission Statement, the CCSS aim to:

> ...provide a consistent, clear understanding of what students are expected to learn, so teachers and parents know what they need to do to help them. The standards are designed to be robust and relevant to the real world, reflecting the knowledge and skills that our young people need for success in college and careers. With American students fully prepared for the future, our communities will be best positioned to compete successfully in the global economy.

Previously, the NJ ASK Grade 3 test was administered over 3 days, with 2 days for language arts literacy, and one for mathematics. Presently, it is administered over 4 days: 2 for English/language arts, and 2 for mathematics. The mathematics test has three major classes of question:

A. The multiple-choice question—This is a traditional multiple choice question, followed by four choices (lettered A–D). For the multiple-choice questions, students are allowed use of a calculator.

B. Short constructed response question (SCR)—This is a question sometimes described as a "multiple choice question without the choices." That is, it has about the same difficulty as a multiple choice question, but the student must calculate the answer, as opposed to choosing (or some times guessing) among the choices listed. The answer is entered into a blank line. For these questions, students may <u>not</u> use a calculator.

C. Extended Constructed Response Question (ECR)—This is a question that requires several steps, and generally a little more time and thought than the other questions. Students are encouraged to think about the answer, show all work (usually more than the other types of questions). Students are encouraged to draw a diagram or a picture to help them through to an answer to the problem.

## FORMAT OF THE NJ ASK, GRADE 3

The NJ ASK 3 Math test is given over the course of 2 days. There are six parts to the test, and students are given 20 minutes to complete each part, so that students are being tested for 60 minutes each day:

## Day 1:

Part 1—This part has eight SCR questions. For this part of the test, students are *not* allowed to use a calculator.

Part 2—This part contains 11 multiple-choice questions. Students are *not* allowed to use a calculator for some of the questions in this part.

Part 3—This part ends the first day, and has the format of the rest of the test. It has eight multiple-choice questions, and one ECR question. For this part, students *may not* use a calculator.

## Day 2:

Part 4—This part has eight multiple-choice questions, and one ECR question. For this part, students *may use* a calculator.

Part 5—This part has eight multiple-choice questions, and one ECR question. For this part, students may *not* use a calculator.

Part 6—This part has eight multiple-choice questions, and one ECR question. For this part, students may *not* use a calculator.

The only part that students can use a calculator on is part 4. Parts 3–6 have the same format. That is, they have eight multiple-choice questions and one ECR question. Students are given 20 minutes to work on parts 1 and 4, 19 minutes to work on part 2, and 24 minutes for parts 3, 5, and 6. They are tested 60 minutes on each of the 2 days of testing.

The NJ CCSS in mathematics cover five main topics: Operations and Algebraic Thinking, Numbers and Operations in Base Ten, Numbers and Operations—Fractions, Measurement and Data, and Geometry. The full, detailed list of topics follows:

## COMMON CORE STATE STANDARDS IN MATHEMATICS FOR GRADE 3

### I. CONTENT

#### Operations and Algebraic Thinking (3.OA)

**3.OA.1.** Interpret products of whole numbers, e.g., interpret $5 \times 7$ as the total number of objects in five groups of seven objects each. *For example, describe a context in which a total number of objects can be expressed as $5 \times 7$.*

**3.OA.2.** Interpret whole-number quotients of whole numbers, e.g., interpret $56 \div 8$ as the number of objects in each share when 56 objects are partitioned equally into eight shares, or as a number of shares when 56 objects are partitioned into equal shares of eight objects each. *For example, describe a context in which a number of shares or a number of groups can be expressed as $56 \div 8$.*

**3.OA.3.** Use multiplication and division within 100 to solve word problems in situations involving equal groups, arrays, and measurement quantities, e.g., by using drawings and equations with a symbol for the unknown number to represent the problem.

**3.OA.4.** Determine the unknown whole number in a multiplication or division equation relating three whole numbers. *For example, determine the unknown number that makes the equation true in each of the equations $8 \times ? = 48$, $5 = ? \div 3$, $6 \times 6 = ?$.*

**3.OA.5.** Apply properties of operations as strategies to multiply and divide. *Examples: If 6 × 4 = 24 is known, then 4 × 6 = 24 is also known (commutative property of multiplication). 3 × 5 × 2 can be found by 3 × 5 = 15, then 15 × 2 = 30, or by 5 × 2 = 10, then 3 × 10 = 30 (associative property of multiplication). Knowing that 8 × 5 = 40 and 8 × 2 = 16, one can find 8 × 7 as 8 × (5 + 2) = (8 × 5) + (8 × 2) = 40 + 16 = 56 (distributive property).*

**3.OA.6.** Understand division as an unknown-factor problem. *For example, find 32 ÷ 8 by finding the number that makes 32 when multiplied by 8.*

**3.OA.7.** Fluently multiply and divide within 100, using strategies such as the relationship between multiplication and division (e.g., knowing that 8 × 5 = 40, one knows 40 ÷ 5 = 8) or properties of operations. By the end of grade 3, know from memory all products of two one-digit numbers.

**3.OA.8.** Solve two-step word problems using the four operations. Represent these problems using equations with a letter standing for the unknown quantity. Assess the reasonableness of answers using mental computation and estimation strategies including rounding.

**3.OA.9.** Identify arithmetic patterns (including patterns in the addition table or multiplication table), and explain them using properties of operations. *For example, observe that 4 times a number is always even, and explain why 4 times a number can be decomposed into two equal addends.*

## Number and Operations in Base Ten (3.NBT)

**3.NBT.1.** Use place value understanding to round whole numbers to the nearest 10 or 100.

**3.NBT 2.** Fluently add and subtract within 1,000 using strategies and algorithms based on place value, properties of operations, and/or the relationship between addition and subtraction.

**3.NBT.3.** Multiply one-digit whole numbers by multiples of 10 in the range 10–90 (e.g., $9 \times 80$, $5 \times 60$) using strategies based on place value and properties of operations.

## Number and Operations—Fractions (3.NF)

["Grade 3 expectations in this domain are limited to fractions with denominators 2, 3, 4, 6, and 8" *(footnote to Common Core State Standards).*]

**3.NF.1.** Understand a fraction $1/b$ as the quantity formed by 1 part when a whole is partitioned into $b$ equal parts; understand a fraction $a/b$ as the quantity formed by $a$ parts of size $1/b$.

**3.NF.2.** Understand a fraction as a number on the number line; represent fractions on a number line diagram.

a. Represent a fraction $1/b$ on a number line diagram by defining the interval from 0 to 1 as the whole and partitioning it into $b$ equal parts. Recognize that each part has size $1/b$ and that the endpoint of the part based at 0 locates the number $1/b$ on the number line.

b. Represent a fraction $a/b$ on a number line diagram by marking off $a$ lengths $1/b$ from 0. Recognize that the resulting interval has size $a/b$ and that its endpoint locates the number $a/b$ on the number line.

**3.NF.3.** Explain equivalence of fractions in special cases, and compare fractions by reasoning about their size.

a. Understand two fractions as equivalent (equal) if they are the same size, or the same point on a number line.
b. Recognize and generate simple equivalent fractions, e.g., 1/2 = 2/4, 4/6 = 2/3). Explain why the fractions are equivalent, e.g., by using a visual fraction model.
c. Express whole numbers as fractions, and recognize fractions that are equivalent to whole numbers. *Examples: Express 3 in the form 3 = 3/1; recognize that 6/1 = 6; locate 4/4 and 1 at the same point of a number line diagram.*
d. Compare two fractions with the same numerator or the same denominator by reasoning about their size. Recognize that comparisons are valid only when the two fractions refer to the same whole. Record the results of comparisons with the symbols >, =, or <, and justify the conclusions, e.g., by using a visual fraction model.

## Measurement and Data (3.MD)

**3.MD.1.** Tell and write time to the nearest minute and measure time intervals in minutes. Solve word problems involving addition and subtraction of time intervals in minutes, e.g., by representing the problem on a number line diagram.

**3.MD.2.** Measure and estimate liquid volumes and masses of objects using standard units of grams (g), kilograms (kg), and liters (l). Add, subtract, multiply, or divide to solve one-step word problems involving masses or volumes that are given in the same units, e.g., by using drawings (such as a beaker with a measurement scale) to represent the problem.

**3.MD.3.** Draw a scaled picture graph and a scaled bar graph to represent a data set with several categories. Solve one- and two-step "how many more" and "how many less" problems using information presented in scaled bar graphs. *For example, draw a bar graph in which each square in the bar graph might represent five pets.*

**3.MD.4.** Generate measurement data by measuring lengths using rulers marked with halves and fourths of an inch. Show the data by making a line plot, where the horizontal scale is marked off in appropriate units—whole numbers, halves, or quarters.

**3.MD.5.** Recognize area as an attribute of plane figures and understand concepts of area measurement.

a. A square with side length one unit, called "a unit square," is said to have "one square unit" of area, and can be used to measure area.
b. A plane figure that can be covered without gaps or overlaps by $n$ unit squares is said to have an area of $n$ square units.

**3.MD.6.** Measure areas by counting unit squares (square centimeter, square meter, square inch, square foot, and improvised units).

**3.MD.7.** Relate area to the operations of multiplication and addition.

a. Find the area of a rectangle with whole-number side lengths by tiling it, and show that the area is the same as would be found by multiplying the side lengths.
b. Multiply side lengths to find areas of rectangles with whole-number side lengths in the context of solving real-world and mathematical problems, and represent whole-number products as rectangular areas in mathematical reasoning.
c. Use tiling to show in a concrete case that the area of a rectangle with whole-number side lengths $a$ and $b + c$

is the sum of $a \times b$ and $a \times c$. Use area models to represent the distributive property in mathematical reasoning.

d. Recognize area as additive. Find areas of rectilinear figures by decomposing them into non-overlapping rectangles and adding the areas of the non-overlapping parts, applying this technique to solve real-world problems.

**3.MD.8.** Solve real-world and mathematical problems involving perimeters of polygons, including finding the perimeter given the side lengths, finding an unknown side length, and exhibiting rectangles with the same perimeter and different areas or with the same area and different perimeters.

## Geometry (3.G)

**3.G.1.** Understand that shapes in different categories (e.g., rhombuses, rectangles, and others) may share attributes (e.g., having four sides), and that the shared attributes can define a larger category (e.g., quadrilaterals). Recognize rhombuses, rectangles, and squares as examples of quadrilaterals, and draw examples of quadrilaterals that do not belong to any of these subcategories.

**3.G.2.** Partition shapes into parts with equal areas. Express the area of each part as a unit fraction of the whole. *For example, partition a shape into four parts with equal area, and describe the area of each part as 1/4 of the area of the shape.*

Now, if you compare these with the previous standards governing the NJ ASK (that is, the NJ CCCS in mathematics) you will note that these new standards are covering fewer standards. This is the intention of the new CCSS: to be more targeted, more focused from one grade to the next, and cover fewer standards per grade. It is believed that if students can concentrate on a smaller number of standards and skills from one grade to the next through elementary and middle school, they will retain more mathematical skills and knowledge by the time they reach high school. In addition to the content standards, there are eight standards for mathematical practice, which follow.

## II. MATHEMATICAL PRACTICE

1. **Make sense of problems and persevere in solving them**—Mathematically proficient students start by explaining to themselves the meaning of a problem and looking for entry points to its solution. They analyze givens, constraints, relationships, and goals. They make conjectures about the form and meaning of the solution and plan a solution pathway rather than simply jumping into a solution attempt. They consider analogous problems, and try special cases and simpler forms of the original problem to gain insight into its solution. They monitor and evaluate their progress and change course if necessary. Older students might, depending on the context of the problem, transform algebraic expressions or change the viewing window on their graphing calculator to get the information they need. Mathematically proficient students can explain correspondences between equations, verbal descriptions, tables, and graphs or draw diagrams of important features and relationships, graph data, and search for regularity or trends. Younger students might rely on using concrete objects or pictures to help conceptualize and solve a problem. Mathematically proficient students check their answers to problems using a different method, and they continually ask themselves, "Does this make sense?" They can understand the approaches of others to solving complex problems and identify correspondences between different approaches.

2. **Reason abstractly and quantitatively**—Mathematically proficient students make sense of quantities and their

relationships in problem situations. They bring two complementary abilities to bear on problems involving quantitative relationships: the ability to *decontextualize*—to abstract a given situation and represent it symbolically and manipulate the representing symbols as if they have a life of their own, without necessarily attending to their referents—and the ability to *contextualize*, to pause as needed during the manipulation process to probe into the referents for the symbols involved. Quantitative reasoning entails habits of creating a coherent representation of the problem at hand; considering the units involved; attending to the meaning of quantities, not just how to compute them; and knowing and flexibly using different properties of operations and objects.

3. **Construct viable arguments and critique the reasoning of others**—Mathematically proficient students understand and use stated assumptions, definitions, and previously established results in constructing arguments. They make conjectures and build a logical progression of statements to explore the truth of their conjectures. They are able to analyze situations by breaking them into cases, and can recognize and use counterexamples. They justify their conclusions, communicate them to others, and respond to the arguments of others. They reason inductively about data, making plausible arguments that take into account the context from which the data arose. Mathematically proficient students are also able to compare the effectiveness of two plausible arguments, distinguish correct logic or reasoning from that which is flawed, and—if there is a flaw in an argument—explain what it is. Elementary students can construct arguments using concrete referents such as objects, drawings, diagrams, and actions. Such arguments can make sense and be correct, even though they are not generalized or made formal until later grades. Later, students learn to determine domains to which an argument applies. Students at all grades can listen or

read the arguments of others, decide whether they make sense, and ask useful questions to clarify or improve the arguments.

4. **Model with mathematics**—Mathematically proficient students can apply the mathematics they know to solve problems arising in everyday life, society, and the workplace. In early grades, this might be as simple as writing an addition equation to describe a situation. In middle grades, a student might apply proportional reasoning to plan a school event or analyze a problem in the community. By high school, a student might use geometry to solve a design problem or use a function to describe how one quantity of interest depends on another. Mathematically proficient students who can apply what they know are comfortable making assumptions and approximations to simplify a complicated situation, realizing that these may need revision later. They are able to identify important quantities in a practical situation and map their relationships using such tools as diagrams, two-way tables, graphs, flowcharts, and formulas. They can analyze those relationships mathematically to draw conclusions. They routinely interpret their mathematical results in the context of the situation and reflect on whether the results make sense, possibly improving the model if it has not served its purpose.

5. **Use appropriate tools strategically**—Mathematically proficient students consider the available tools when solving a mathematical problem. These tools might include pencil and paper, concrete models, a ruler, a protractor, a calculator, a spreadsheet, a computer algebra system, a statistical package, or dynamic geometry software. Proficient students are sufficiently familiar with tools appropriate for their grade or course to make sound decisions about when each of these tools might be helpful, recognizing both the insights to be gained and their limitations. For example, mathematically proficient high school students analyze graphs of functions and solutions

generated using a graphing calculator. They detect possible errors by strategically using estimation and other mathematical knowledge. When making mathematical models, they know that technology can enable them to visualize the results of varying assumptions, explore consequences, and compare predictions with data. Mathematically proficient students at various grade levels are able to identify relevant external mathematical resources, such as digital content located on a website, and use them to pose or solve problems. They are able to use technological tools to explore and deepen their understanding of concepts.

6. **Attend to precision**—Mathematically proficient students try to communicate precisely to others. They try to use clear definitions in discussion with others and in their own reasoning. They state the meaning of the symbols they choose, including using the equal sign consistently and appropriately. They are careful about specifying units of measure, and labeling axes to clarify the correspondence with quantities in a problem. They calculate accurately and efficiently, and express numerical answers with a degree of precision appropriate for the problem context. In the elementary grades, students give carefully formulated explanations to each other. By the time they reach high school, they have learned to examine claims and make explicit use of definitions.

7. **Look for and make use of structure**—Mathematically proficient students look closely to discern a pattern or structure. Young students, for example, might notice that three and seven more is the same amount as seven and three more, or they may sort a collection of shapes according to how many sides the shapes have. Later, students will see $7 \times 8$ equals the well remembered $7 \times 5 + 7 \times 3$, in preparation for learning about the distributive property. In the expression $x^2 + 9x + 14$, older students can see the 14 as $2 \times 7$ and the 9 as $2 + 7$. They recognize the significance of an existing

line in a geometric figure and can use the strategy of drawing an auxiliary line for solving problems. They also can step back for an overview and shift perspective. They can see complicated things, such as some algebraic expressions, as single objects or as being composed of several objects. For example, they can see $5 - 3(x - y)^2$ as 5 minus a positive number times a square and use that to realize that its value cannot be more than 5 for any real numbers $x$ and $y$.

8. **Look for and express regularity in repeated reasoning**—Mathematically proficient students notice whether calculations are repeated, and look both for general methods and for shortcuts. Upper elementary students might notice when dividing 25 by 11 that they are repeating the same calculations over and over again, and conclude they have a repeating decimal. By paying attention to the calculation of slope as they repeatedly check whether points are on the line through (1, 2) with slope 3, middle school students might abstract the equation $(y - 2)/(x - 1) = 3$. Noticing the regularity in the way terms cancel when expanding $(x - 1)(x + 1)$, $(x - 1)(x^2 + x + 1)$, and $(x - 1)(x^3 + x^2 + x + 1)$ might lead them to the general formula for the sum of a geometric series. As they work to solve a problem, mathematically proficient students maintain oversight of the process, while attending to the details. They continually evaluate the reasonableness of their intermediate results.

# OPERATIONS AND ALGEBRAIC THINKING

We use numbers every day in everything we do. We need to find out how hot or cold it is on a particular day, so we can put on the right article of clothing, and we determine how big or small those clothes are with numbers. We measure the air in our tires and the cost of the gas in our tank using numbers, among many other tasks. Our first standard in third grade deals with multiplication and division.

## MULTIPLICATION

Multiplication is a process very similar to addition. In fact, it is addition on an extended scale. Also like addition, multiplication has a set of basic facts that must be memorized. These basic facts are inclusive of the factors through 10. This means that you should memorize and practice all multiplication facts that are up to $10 \times 10$. Memorizing these multiplication facts is very important because a good portion of third grade, and basically every grade after that, will rely on these as a foundation for doing other larger and more complicated problems. If you do not know the facts, you will face great difficulty solving many future math problems.

## MULTIPLICATION FACTS

There are a few tricks to memorizing the facts. Not every numeral has a trick to it; some just require memorization on your part. We'll go through a few of the tricks here.

- For all facts, you can do skip counting. Skip counting is the process by which you count by the given number. For example, if I asked you what $5 \times 3$ is, you could count by 5s three times. 5, 10, 15; so 15 is the answer. You could also do it backwards if that is easier for you; count by 3s five times. 3, 6, 9, 12, 15; and again we see that 15 is the answer.

- Similar to skip counting is multiplication using arrays. For example, if you were multiplying $4 \times 7$ you would make four groups of seven (using dots, or counters, or any other appropriate manipulative) and add them up to get a total of 28. An example would be an array for $4 \times 7$ that shows four columns of seven dots in each or seven columns of four dots in each.

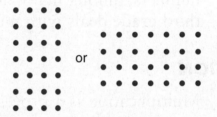

## Other Tricks for Memorizing Multiplication Facts

- Any number times 0 is 0 (for example, $8 \times 0 = 0$).
- Any number times 1 is itself (for example, $9 \times 1 = 9$).
- Any number times 2 is double itself (for example, $6 + 6 = 12$ which is the same as $6 \times 2$).
- Any number times 5 will either end in 5 or 0 (for example, $6 \times 5 = 30$, and $5 \times 7 = 35$). How does this help? Well if you come up with an answer that ends in anything other than a 5 or a 0, you know it is incorrect. If you have a multiple choice test and some of the choices end in something other than 5 or 0, you can cross them off because they are wrong. This helps narrow down the actual answer.
- $6 \times 8 = 48$ (this one rhymes; six times eight is forty-eight).

■ There is a neat trick for all 9s up to 9 × 10. Let's say we're multiplying 9 × 7. Hold both of your hands in the air and spread apart all 10 of your fingers (when we refer to fingers, we are including thumbs). This trick *only* works for the 9s facts. Since we're doing 9 × 7, start at your left pinky and count in your head to your seventh finger. You should end at your right index finger (next to your thumb). Put that finger down and keep all the others up. Now, look how many fingers are to the left of the finger you just put down. You should see six. Now look how many fingers are to the right of the finger you just put down. You should see three. Put those two numbers (6 and 3) together and you have your answer to 9 × 7 = 63. This trick works with any single-digit number when multiplied by 9.

■ There is also a trick you can do by rounding up the number then subtracting backwards. For example, if you did not know that 9 × 7 = 63, you could do 10 × 7, which is a lot easier, and get 70. Then subtract a 7 and get 63. This works because originally we asked how much nine 7s were. If instead you do ten 7s to make it easier, you can just subtract a 7 afterwards and still get the correct answer. Here's another example of it. 5 × 9 = 45. If you didn't know this immediately, and could not figure it out by skip counting, you could instead do 5 × 10 and get 50. Then just subtract a 5 (because you did ten 5s instead of nine 5s) and get 45.

Don't forget that multiplication facts can be reversed. 4 × 7 is the same as 7 × 4. This is known as the commutative property of multiplication. It is an identical process to turnaround facts in addition (3 + 4 is the same as 4 + 3).

## Multiplication Fact Exercises

Answers appear on page 183.

1. What is 5 × 7?

   **A.** 35

   **B.** 40

   **C.** 30

   **D.** 23

2. What is 9 × 2?

   **A.** 16

   **B.** 18

   **C.** 13

   **D.** 9

3. What is 9 × 7?

   **A.** 63

   **B.** 61

   **C.** 72

   **D.** 54

4. What is 7 × 7?

   **A.** 42

   **B.** 49

   **C.** 7

   **D.** 1

5. What is 8 × 0?

   **A.** 8

   **B.** 1

   **C.** 0

   **D.** 16

## DIVISION

Division is the process of splitting a number into smaller pieces. For example, if you are given the number 15 and asked to split it into five equal pieces, the answer would be 3. You could make five equal pieces with three in each, totaling 15. Division involves grouping, or equally breaking apart a given number into smaller groups. It can be solved using an elementary method by using counters, or coins. The down side of this method is that it can take a very long time to count out coins, split them into equal groups, and hope that you did not drop one on the floor by accident or miscount to begin with. Also, you have to be certain you do not give any groups any more than the other groups have (they must be equal). This can also be challenging. Division should be thought of as the opposite of multiplication. If you are good at multiplication facts, you will probably be good at division facts. To solve smaller division problems, just read them backwards as a multiplication problem.

Another way to compute a division problem is by skip counting. If you were asked the question: $10 \div 2 = ?$, you could solve this pretty easily with skip counting. Think of it this way, how many times do you have to count by 2s until you reach ten? If you counted aloud, or in your head, you would see that you need to count by 2s five times to reach 10. Hence, the answer is 5. This method only really works with very small division problems. It becomes difficult to skip count if you are asked a question such as $49 \div 7 = ?$ because it can be confusing to count by 7s. In this case, you would just need to read it backwards as described before and realize that $7 \times 7 = 49$, so the answer is 7. There are seven 7s in 49. Alternately, you could take a piece of scrap paper, make seven circles on it, and one-at-a-time put a dot in each circle until you reach 49. Just make sure you have an equal number of dots in each circle. When finished, you would merely count up how many dots were in each circle and *presto*, you have your answer. Here is what it might look like:

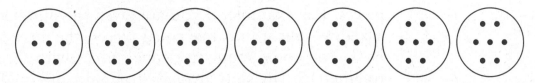

As you can imagine, this method also leaves a lot of room for error, but it is important to realize it is a backstop, or an alternative, in case you have no other reasonable means to figure out the answer.

In third grade, it is required that students learn all of the division facts, similar to that of the multiplication facts within 100. The process of long division begins in fourth grade where students will hand write division problems using a longer method (because the division problems will be much larger). This may also include remainders where you cannot divide into equal groups. So at this time, it is important for students to memorize those multiplication facts, because this way they will also know their division facts just as well.

## Division Fact Exercises

Answers appear on pages 183 and 184.

1. What is 42 ÷ 6?

    **A.**   6

    **B.**   7

    **C.**  12

    **D.** 252

2. What is 25 ÷ 5?

    **A.**   5

    **B.** 125

    **C.**   3

    **D.**   4

**3.** What is 63 ÷ 7?

   **A.** 461

   **B.** 8

   **C.** 7

   **D.** 9

**4.** What is 30 ÷ 5?

   **A.** 6

   **B.** 15

   **C.** 5

   **D.** 2

**5.** What is 36 ÷ 9?

   **A.** 2

   **B.** 8

   **C.** 4

   **D.** 7

## Multiplication and Division Fact Families

Division and multiplication are directly related in the same way addition and subtraction are. They are so related that you can create fact families of four problems (two division and two multiplication) using just three numbers (exactly as you can do with addition and subtraction). In a fact triangle such as this, you would have two multiplication problems. Each problem has two factors and one product. The factors are the numbers being multiplied together, and the product is the answer. There are also two division problems to be created. Each would consist of a dividend (the first number in the division problem, and also the largest of the three numbers), a divisor (either of the two lesser numbers), and quotient (the answer). For example, shown graphically as a fact triangle, it would look like this:

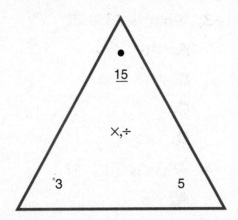

From this we can derive:

$$3 \times 5 = 15$$

$$5 \times 3 = 15$$

$$15 \div 3 = 5$$

$$15 \div 5 = 3$$

It is important to note that on fact triangles, the number on top has a dot just above it. For your two multiplication problems, this is your product (the answer when you multiply the two other numbers). For your two division problems, this is your dividend (the number that goes first in each of the two division problems.) In this type of fact triangle, any of the values may be missing, and it would be your responsibility to determine the correct value to insert, so it is important that you understand the meaning of each value's location on the fact triangle. Furthermore, on an exam, you may be asked to create a fact family for a missing variable. For example:

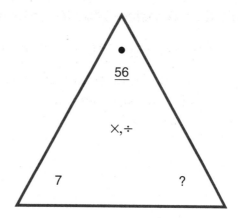

If shown the fact triangle listed above, you begin by solving for the missing variable (note that on an exam you may see the missing variable listed as a ?, or a manuscript letter such as $x$, $r$, $b$, etc.). This can be solved easily and the solution can be further extended by creating the four facts that go along with this triangle.

$$56 \div 7 = ?$$

$$56 \div ? = 7$$

$$7 \times ? = 56$$

$$? \times 7 = 56$$

If you know your multiplication facts well, you will realize the answer is 8. The division process here is taking the number 56, splitting it into seven equal groups, then realizing that to do that, you would have exactly eight in each group with none left over. Recalling your multiplication facts provides the best and quickest way to come up with a solution. The takeaway from this exercise is understanding the direct relationship between multiplication and division. This includes solving for a missing variable, and creating a fact family.

## Missing Variable and Division/Multiplication Fact Family Exercises

Answers appear on pages 184 and 185.

1. In the fact triangle shown below, what is the value of the missing variable?

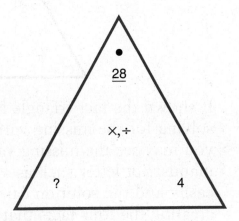

A. 3

B. 7

C. 4

D. 148

**2.** In the fact triangle shown below, what is the value of the missing variable?

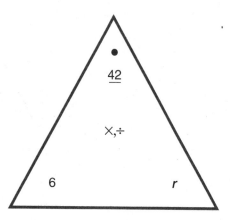

**A.** 7

**B.** 6

**C.** 4

**D.** 8

**3.** In the fact triangle shown below, what is the value of the missing variable?

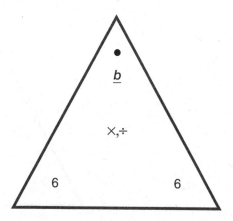

**A.** 1

**B.** 12

**C.** 0

**D.** 36

**4.** In the fact triangle shown below, what is the value of the missing variable?

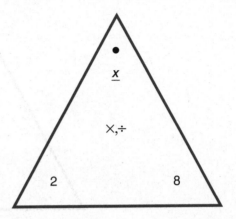

**A.** 2

**B.** 3

**C.** 16

**D.** 12

**5.** In the fact triangle shown below, what is the value of the missing variable?

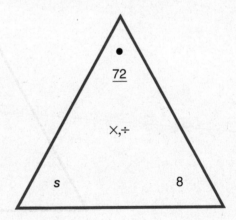

**A.** 9

**B.** 8

**C.** 11

**D.** 6

**6.** What is the fact family for the multiplication/division fact triangle shown below?

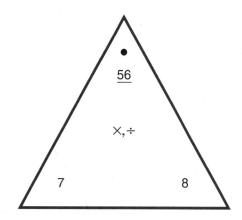

Answer:  _____ × _____ = _____

_____ × _____ = _____

_____ ÷ _____ = _____

_____ ÷ _____ = _____

**7.** What is the fact family for the multiplication/division fact triangle shown below?

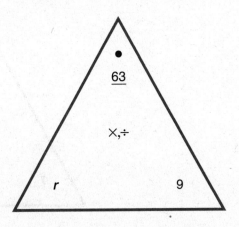

Answer:   _____ × _____ = _____

_____ × _____ = _____

_____ ÷ _____ = _____

_____ ÷ _____ = _____

**8.** What is the fact family for the multiplication/division fact triangle shown below?

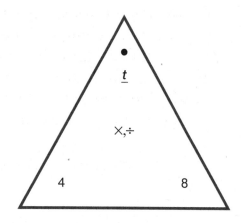

Answer: _____ × _____ = _____

_____ × _____ = _____

_____ ÷ _____ = _____

_____ ÷ _____ = _____

## SINGLE-STEP WORD PROBLEMS

Now that we have a complete understanding of multiplication and division, we can further expand this to solve word problems within 100. Solving word problems requires critical thinking techniques to identify important information and how it relates to the missing variable. It is vital that you read each problem slowly and carefully so you do not choose an incorrect operation to solve it.

**Let's try a practice problem:**

Mrs. Nicosia just purchased nine boxes of crayons to replace some broken ones in her classroom. Each box of crayons holds eight crayons. How many crayons in all did she purchase? The first step in tackling this problem is identifying the information given; then we can look toward figuring out what to do to compute the missing variable. To organize our work, we will fill in a table:

| Boxes of crayons | Crayons per box | Crayons in all |
|:---:|:---:|:---:|
| 9 | 8 | ? |

Answer: _____
(unit)

Now that our information is organized and easy to view, we can proceed to solve. In this word problem, we can see that we will have to use multiplication to solve it. If each box has eight crayons in it, we have to multiply that by the nine boxes that were purchased. Another way to look at it would be to count by 8s nine times. Therefore, the correct answer in this problem is 72 crayons (remember to include the unit "crayons" in your answer).

Here are some practice problems for you to try. Answers appear on pages 186 to 188.

1. Hilga has 24 books, in four equal piles. How many books are in each pile?

| Piles | Books per pile | Books in all |
|:---:|:---:|:---:|
| | | |

Answer: _____
(unit)

**2.** Wally sees a full parking lot that has six rows of cars, and each row has eight cars in it. How many cars does Wally see in this lot?

| Rows | Cars per row | Cars in all |
|------|-------------|-------------|
|      |             |             |

Answer: _____
(unit)

**3.** Jamie teaches gymnastics. She teaches three classes each day, 4 days per week. In one week, how many classes will Jamie have taught?

| Days | Classes per day | Classes in all |
|------|----------------|----------------|
|      |                |                |

Answer: _____
(unit)

**4.** Jennifer bought tacos for her seven friends and herself. She bought two tacos for herself and two tacos for each of her friends. How many tacos did she buy?

| People | Tacos per person | Tacos in all |
|--------|-----------------|--------------|
|        |                 |              |

Answer: _____
(unit)

**5.** Teddy's 10th grade class is going to the Bluewater
Aquarium. The school has vans to get the students
there. Each van holds 6 students, and there are 24
students in the class. How many vans will be needed
to drive the students to the aquarium?

| Vans | Students per van | Students in all |
|------|------------------|-----------------|
|      |                  |                 |

Answer: _____
(unit)

## TWO-STEP WORD PROBLEMS

Two-step word problems are slightly more advanced than
the single-step problems previously discussed. The
difference is implied in the name. Two-step word
problems require you to do some mathematical operation
before you can begin solving the problem. You may have
to add some numbers up, divide two numbers, etc. The
level of critical thinking is more advanced also because it
will require you to determine what combination of the
four operations ($\times$, $\div$, $+$, $-$) you will need to utilize. Let's
take a look at an example:

Mr. Jenkins travels a good distance to work every day.
In his work commute, Mr. Jenkins's car uses two gallons
of gas per day. If he works 5 days per week, then how
many gallons of gas would he use per month (4 weeks)
due to his work?

The answer to this problem will require two
mathematical steps. First, if the car uses two gallons per
day, we need to figure out how many gallons it uses per
week. The problem states he works 5 days per week, so
we multiply $2 \times 5$ and get 10. Each week his car uses 10
gallons of gas because of work. Now that we know the
weekly total for gas, we can multiply this by 4 because
there are 4 weeks in a month. $10 \times 4 = 40$. So each
month, the car uses 40 gallons of gas due to Mr. Jenkins's

work. This problem required multiplication in both steps, but it should be noted that it could realistically be any combination of the four mathematical operations.

Here are some practice problems for you to try. Answers appear on pages 188 and 189.

1. Ron is buying some fresh fruit. He bought 6 oranges, 3 grapefruits, 11 lemons, and 20 limes. If each piece of fruit costs $0.10, how much did the fruit cost in all?

    Answer: _____

2. Marilyn likes to walk her grandson Mark to the park for a stroll. He drinks one bottle of water every hour. If they will be there for 2 hours, how many bottles of water will Marilyn need to purchase each week (7 days) to take to the park?

    Answer: _____

3. Carly babysits for her next door neighbor. Last month, she babysat for seven nights and sat the following number of hours: 2 hours, 3 hours, 2 hours, 4 hours, 5 hours, 3 hours, and 1 hour. If she is paid $6.00 per hour, how much money did she make babysitting for her next door neighbor last month?

    Answer: _____

4. Tom cuts lawns for a little cash. If he cuts three lawns per week, and works 3 hours on one, 2 hours on a second one, and 4 hours on a third one, how many hours will he have worked in a month (4 weeks)?

    Answer: _____

**5.** Yanni went to the video store, and bought 6 DVDs, 11 CDs, and 7 Blu Ray discs. When he got home, he realized he already had two of the DVDs he just purchased. So as to not waste money, Yannie goes back to the store and returns those two DVDs. After returning those two DVDs, how many items did Yanni end up purchasing?

Answer: _____

## PATTERNS

Patterns are all around us. Many of the nursery rhymes that you are familiar with describe patterns.

*Hickory, dickory, dock,*

*A mouse ran up the clock,*

*The clock struck one,*

*And down he run,*

*Hickory, dickory dock.*

It has been suggested that the words hickory, dickory, and dock stand for "eight," "nine," and "ten," respectively. In the schoolyard verse for choosing sides, "eeny, meeny, miny, mo" most likely stand for: "one, two, three, four."

The familiar nursery rhyme:

*As I was going to St. Ives,*

*I met a man with seven wives,*

*And every wife had seven sacks,*

*And every sack had seven cats,*

*And every cat had seven kits,*

*Kits, cats, sacks, wives,*

*How many were going to St. Ives?*

… simply suggests a multiplication problem: $7 \times 7 \times 7 \times 7$.

## ACTIVITY: LET'S EXPLORE A PATTERN

Get out a piece of paper. Write 7 up at the top. Add 5 to the 7 and put the result below the 7. Add 5 to the result and put the result below. Continue to do this until you get to around 120. Now, look at the sequence of numbers.

Do you notice any patterns? Are there any patterns in the units place? In the tens place? Add two successive numbers. Do you see a pattern?

Mathematics is full of patterns. Some patterns are easy to see, such as the even numbers (2, 4, 6, 8, etc.). Some are not so easy to see, such as the Fibonnacci sequence (1, 1, 2, 3, 5, 8, 13, 21, etc.). In case you did not see it, in the Fibonnacci sequence, the next number is found by adding the previous two numbers together (try it with the numbers of the sequence yourself). Mathematics is considered to be the search for patterns in our world. Formulas, such as the area of a rectangle ($A = L \times W$), or area is length times width, is nothing more than a pattern for finding how big a rectangle is. Using patterns, we can find out how big a wall is, or how big a room is (volume). Using patterns, we can predict (to the second) when the sun will rise and when it will set.

Number arrays (as discussed earlier) can have patterns. The array of numbers:

$$2, 4, 6, 8, 10, 12\ldots$$

is probably familiar to you. It is the even numbers. Now, look at this pattern:

$$1, 3, 5, 7\ldots$$

What would the next three numbers in this array be? The next three numbers in this array are 9, 11, and 13, because these are the odd numbers.

You can count down and still find patterns in numbers. Look at this array:

$$55, 51, 47, 43, 39\ldots$$

What would the next three numbers be? The next numbers would be 35, 31, and 27 because you are counting down by 4s.

Similarly, patterns can be represented in a table, as seen on the next few pages. In this type of representation you have three components: the rule, the "in" number, and the "out" number. In the example below, we are given the rule of ×2. For those "in" numbers we have, we can simply apply the rule and obtain the "out" number.

| RULE |
|------|
| ×2 |

| IN | OUT |
|----|-----|
| 6 | |
| 7 | |
| | 18 |
| | 6 |
| 4 | |

So 6 × 2 = 12, 7 × 2 = 14, and 4 × 2 = 8. For those locations where we are missing the "in" number, we must consider what number ×2 equals the "out" number. Alternately, we could think of this as a division problem: the "out" number divided by the rule. So 18 ÷ 2 = 9 and 6 ÷ 2 = 3.

| RULE |
|------|
| ×2 |

| IN | OUT |
|----|-----|
| 6 | 12 |
| 7 | 14 |
| 9 | 18 |
| 3 | 6 |
| 4 | 8 |

The concept of patterns can be further extended in this type of table by giving all variables in the table and requiring you to determine the rule. In an example such as this, it is important to compare multiple rows of numbers to be certain the rule is what it appears to be.

| RULE |
|------|
|      |

| IN | OUT |
|----|-----|
| 41 | 47 |
| 101 | 107 |
| 3 | 9 |
| 14 | 20 |

In an example such as this one, we take one of the "in" numbers and compare it to the "out" number. If we compare 41 to 47, we can see it increased by 6, so the rule should be +6. However, to be certain, always check another row. If we move down and try the 3 and 9, we can see that the rule of +6 does in fact work.

## Patterns Practice

Answers appear on pages 189 to 192.

**1.** What are the next four numbers in the number array: 5, 8, 11, 14,...

_____  _____  _____  _____

Explain the pattern:

_____

**2.** What are the next three numbers in the number array: 256, 251, 245, 238,...

_____  _____  _____

Explain the pattern:

_____

**3.** Complete the table below and apply the rule to make your own "in" and "out" combination in the bottom row of the table:

| RULE |
|------|
| ×6 |

| IN | OUT |
|-----|-----|
| 6 | |
| 4 | |
| | 30 |
| | 6 |
| | |

**4.** Complete the table below and apply the rule to make your own "in" and "out" combination in the bottom row of the table:

| RULE |
|------|
| ×4 |

| IN | OUT |
|-----|-----|
| | 28 |
| | 4 |
| 9 | |
| 0 | |
| | |

**5.** Complete the table below by finding the value for the "rule," then apply the rule to make your own "in" and "out" combination in the bottom row of the table:

| RULE |
|------|
| |

| IN | OUT |
|-----|-----|
| 3 | 6 |
| 9 | 18 |
| 22 | 44 |
| 14 | 28 |
| | |

6. Complete the table below and apply the rule to make your own "in" and "out" combination in the bottom row of the table:

| RULE |
|------|
| ÷2 |

| IN | OUT |
|-----|-----|
| 6 | |
| | 8 |
| | 5 |
| 20 | |
| | |

7. Complete the table below and apply the rule to make your own "in" and "out" combination in the bottom row of the table:

| RULE |
|------|
| +4 |

| IN | OUT |
|-----|-----|
| | 7 |
| 28 | |
| | 47 |
| | 61 |
| | |

# NUMBERS AND OPERATIONS IN BASE TEN

Numbers are all around us. We use and see numbers on a daily basis. Numbers can be used to represent given values and they can be used in mathematical operations as well. In the grocery store alone, there are many different uses of numbers: weights of items, cost of items, aisles in the store, quantity in a package, and many more. Also, in the grocery store it can become quite difficult to keep track of an exact cost for your groceries. To simplify this, we use a process called estimation, or rounding. To understand the process of rounding, we must first discuss place value concepts.

## DEFINITIONS: PLACE VALUE, ESTIMATION

Every number has what is called a **place value**. The place value of a number is determined by its location in relation to the decimal point. For estimation purposes, we need to understand rounding whole numbers to the nearest 10, 100, or 1,000.

**Estimation** is the process by which one can round a number to a different number that is more easily computed using mental math.

## WHOLE NUMBER PLACE VALUE

**5,461**

The number above is read as five thousand four hundred sixty one. Note that the word "and" was not used anywhere in the extended notation. The word "and" is commonly said out loud when repeating a long number such as this (and is often inserted into the sentence in place of the comma). However, the word "and" actually refers to a decimal position in a number, not a comma.

In the number 5,461, each digit has a specific place value that is determined by its overall position. The place values are listed as follows:

| Number | 5 | 4 | 6 | 1 |
|---|---|---|---|---|
| Place value | Thousands | Hundreds | Tens | Ones |

To the right of the numeral one is a decimal. If there are no digits to the right of the decimal, the decimal is generally not written. It is assumed to be located at the end of the number. When you read place values from the decimal it begins (going to the left) as ones, tens, hundreds, thousands, ten-thousands, and finally hundred-thousands. When we discuss decimals shortly, the same rule will be used, but this time we will be going to the right of the decimal. In the large whole number 5,461 listed above, the place value of each digit is listed underneath it. The place value is permanent, or fixed. The place value of each digit gives that specific digit a certain value. For example, the 4 is in the hundreds position. Therefore, the 4 is worth 400 because there are four of them in the hundreds position. To prove this, you could count by hundreds four times. At the end you would have an answer of 400. Likewise, the 5 in the thousands position is worth 5,000 because 5 thousands are a total of 5,000.

## WHOLE NUMBER PLACE-VALUE EXERCISES

Answers appear on page 193.

1. In the number 3,091, which digit is located in the hundreds place?

   **A.** 3

   **B.** 0

   **C.** 9

   **D.** 1

2. In the number 3,478, which digit is located in the ones position?

   **A.** 4

   **B.** 3

   **C.** 8

   **D.** 7

3. In the number 8,562, which digit is located in the thousands position?

   **A.** 5

   **B.** 2

   **C.** 6

   **D.** 8

4. In the number 1,037, which digit is located in the thousands position?

   **A.** 1

   **B.** 7

   **C.** 3

   **D.** 0

**5.** In the number 9,471, which digit is located in the hundreds position?

**A.** 7

**B.** 9

**C.** 1

**D.** 4

## ESTIMATING

Estimating is the process by which we round a number to make it easier to compute using mental math. Sometimes estimation is referred to as a "ballpark estimate." It gives us a general or rough idea of the value. This can be useful in many situations, such as when you are in the supermarket. In this case, you would generally not have a calculator handy to compute the cost of the items you are purchasing. Using mental math and estimation, it is possible to tabulate (as you shop) to determine what your total should roughly be.

The estimating process is quite simple, but does follow a few basic rules. Unless noted by the question, or the answers to the question, estimation is not an exact science. Different individuals will estimate numbers to different values as they see fit using mental math.

When estimating a number, if we are told to estimate it to the nearest hundred, we must look at the digit in the hundreds position. Once we have located that, to estimate this number, we must look to the right. If the digit to the right is 5 or greater, we will bump up the number in the hundreds position and make everything else to the right a 0. If the digit is less than 5, we will leave the number in the hundreds position the same and make all other digits to the right into zeros.

For example, in the number 871 the 8 is in the hundreds position. If we were asked to round 871 to the nearest hundred, we would locate the 8 because it is in the hundreds position, then look to the right. The number to the right is 7. Seven is greater than 5, so according to the estimation rule, the 8 must be bumped up to 9. All other digits to the right become 0. So 871, rounded to the nearest hundred, will be 900. This makes sense because 871 is closer to 900 than it is to 800. It is 71 away from 800 (871 − 800 = 71) yet only 29 away from 900 (900 − 871 = 29).

Likewise, if we were given the number 6,435 and we were asked to round it to the nearest thousand, we would do the same thing. Looking at the number, we can see that we have a 6 in the thousands position. To the right of it is a 4. Four is less than 5, so applying the estimation rule, we must leave the 6 alone and make all other numbers to the right into zeros. So 6,435 rounded to the nearest thousand would be 6,000 (6,435 is closer to 6,000 than it is to 7,000).

Taking it a step further, if we have a larger number, for example 42,361, we may be asked to round it to the hundred position as well. This is still a straightforward problem, as the two previous examples were. The only difference is that there are extra digits to the left of the digit being rounded. So let's take a look. If we are asked to round 42,361 to the nearest hundred, we would point out the digit in the hundreds position, which is the 3. Looking to the right we see a 6. Six is greater than 5, so the rule says to bump up the digit in the hundreds and make everything to the right zeros. The other digits to the left of the 3 stay as they are. So 42,361 rounded to the nearest hundred would be 42,400.

## ESTIMATION EXERCISES

Answers appear on pages 193 and 194.

1. What is 587 rounded to the nearest hundred?

   **A.** 500

   **B.** 580

   **C.** 600

   **D.** 590

2. What is 3,971 rounded to the nearest thousand?

   **A.** 3,980

   **B.** 3,900

   **C.** 4,100

   **D.** 4,000

3. Which response below shows a digit that is rounded to the hundreds?

   **A.** 9,640

   **B.** 32,800

   **C.** 5,222

   **D.** 8

4. Which response below shows a digit that is rounded to the thousands?

   **A.** 76,000

   **B.** 87,901

   **C.** 14,814

   **D.** 5,400

5. What is 3,971 rounded to the nearest thousand?

   **A.** 3,900

   **B.** 3,000

   **C.** 3,970

   **D.** 4,000

6. What number, if rounded to the tens, would have an answer of 80?

   **A.** 86

   **B.** 74

   **C.** 84

   **D.** 70

7. What number, if rounded to the hundreds, would have an answer of 900?

    **A.** 944

    **B.** 849

    **C.** 951

    **D.** 989

8. What number, if rounded to the hundreds, would have a 7 in the hundreds position?

    **A.** 1,600

    **B.** 1,632

    **C.** 7,789

    **D.** 1,656

## ADDITION

Addition is the arithmetic process of combining numbers together. It is used in many tasks throughout a normal day. Many times it may not even present itself as an obvious math problem. It may be something you do on a regular basis but never realized you were using math to do it. Mental math may be used to compute many smaller addition problems, and do estimation as previously described. However, it can also become necessary to use paper and pencil or a calculator to solve the larger problems.

Smaller problems such as 9 + 8 (also known as addition facts, because both numbers being added are only one digit) can be done using mental math. These are basic facts that should be memorized to make the larger problems easier to deal with. Using mental math, you should come up with an answer of 17 for 9 + 8.

As the problems grow larger, it will become necessary to use paper and pencil to compute the answer. In the third grade, students should be able to add and subtract fluently within 1,000.

Let's take a look at a two-digit number plus another two-digit number such as 54 + 67. One of the keys to solving addition and also subtraction problems is that you should always rewrite them vertically (up and down). 54 + 67 is written horizontally (side-by-side) and is difficult to solve because the place values are not lined up on top of each other. Rewriting the problem and lining up the place value columns makes this problem much easier to solve.

$$\begin{array}{r} 5\ 4 \\ +\ 6\ 7 \\ \hline \end{array}$$

In any addition problem, you must always start adding on the right side. So in the example shown above, the first two digits we add are the 4 and the 7. This yields an answer of 11. Underneath the 7, we write the first part of our answer, a 1. The 1 represents the numeral in the ones position in the number 11. The other 1 (from the tens position) will get carried over on top of the 5 and added to it.

$$\begin{array}{r} {}^{1}\phantom{00} \\ 5\ 4 \\ +\ 6\ 7 \\ \hline 1 \end{array}$$

Now you are ready to begin adding in the tens column. Here we have the 1 we carried, the 5, and the 6. This gives us a total of 12. Normally we would write the 2 underneath the 6, and carry the 1. However, there is nowhere to carry the 1 to. There are no other digits to the left. If you pretend to carry the 1, you will see that it lands in the hundreds position, which is correct, even though there are no other numerals to add to it. So you can either do this, or just bring it down to the bottom for a final answer of 121.

```
          1
        5  4
     +  6  7
     ──────────
     1  2  1
```

This process remains the same no matter how many digits you are adding. It could be 9,581 + 372. Again, you would rewrite the problem vertically (9,581 on top of the 372). Notice in the figure below that the place values of the two numbers are lined up properly. Everything is pushed as far to the right as possible.

```
     9,  5  8  1
  +      3  7  2
  ─────────────────
```

Now you can begin adding the digits starting in the ones position (all the way to the right), and moving over to the left until there is nothing more to add. Be absolutely certain that when you rewrite the problem you line everything up on the right side.

```
          1
     9,  5  8  1
  +      3  7  2
  ─────────────────
     9,  9  5  3
```

After properly adding all of the place value columns, we can see that our final answer works out to 9,953.

## ADDITION EXERCISES

Answers appear on pages 194 and 195.

**1.** What is 58 + 9?

    **A.** 47

    **B.** 48

    **C.** 67

    **D.** 57

**2.** What is 341 + 509?

    **A.** 850

    **B.** 840

    **C.** 950

    **D.** 805

**3.** Which statement is correct?

    **A.** 413 + 613 = 816

    **B.** 413 + 301 = 714

    **C.** 701 + 161 = 956

    **D.** 192 + 339 = 481

**4.** What is 543 + 399 + 42?

    **A.** 984

    **B.** 945

    **C.** 1,014

    **D.** 694

**5.** What is 41 + 32 + 99 + 14?

    **A.** 192

    **B.** 206

    **C.** 184

    **D.** 186

## SUBTRACTION

Subtraction is the arithmetic process of separating or removing one value from another. Like addition, it is also a process that is commonly used on a daily basis using mental math. Subtraction is only slightly more difficult than addition. The main difference is that borrowing can be involved if the subtraction is not possible (for example, if you are trying to take a larger number from a smaller one, such as $3 - 7$). We will get into that more difficult version shortly.

Subtraction is the opposite of addition, just as multiplication and division are opposites. Whereas you may have the problem $53 + 28 = 81$, you can read it backwards as $81 - 28 = 53$. This problem should be solved using paper and pencil because it will require borrowing. Again, like addition, you should *always* rewrite the problem vertically before attempting to solve it. Many times on exams you will find it written horizontally (sideways, like a number model). It is important to rewrite it vertically, being careful to line up the place values properly.

$$
\begin{array}{r}
8\ \ 1 \\
-\ \ 2\ \ 3 \\
\hline
\end{array}
$$

Just like addition, we line the numbers up on the right side, then begin subtracting on the right side. So we will first compute 1 minus 3. The answer is *not* 2. You always read the problem going down, not up. 1 minus 3 is actually a negative number. You cannot take 3 from 1. Imagine if you had 1 piece of candy, and a friend wanted to take 3 pieces from you. It is not possible. So to fix this, we must borrow. Borrowing is the process where you will take 1 from the neighboring numeral immediately to the left of it. In this case, the next numeral is the 8. So cross out the 8, make it a 7 (one less), and add 10 to the original 1 on the right that you were trying to subtract.

$$\begin{array}{r} {\scriptstyle 7 \quad 11} \\ \cancel{8} \ \cancel{1} \\ - \ 2 \ 3 \\ \hline \end{array}$$

That makes the new subtraction problem in the first column on the right 11 – 3. This would yield an answer of 8, so write 8 underneath the 3.

$$\begin{array}{r} {\scriptstyle 7 \quad 11} \\ \cancel{8} \ \cancel{1} \\ - \ 2 \ 3 \\ \hline 8 \end{array}$$

Now we move to the left. 7 minus 2 is 5. So bring that 5 down.

$$\begin{array}{r} {\scriptstyle 7 \quad 11} \\ \cancel{8} \ \cancel{1} \\ - \ 2 \ 3 \\ \hline 5 \ 8 \end{array}$$

There are no more digits to subtract to the left so our answer of 58 is final.

The borrowing process is by far the most common point of error. You must be very careful to not just borrow from the neighboring digit, without making it one less. Always show your work each step of the way to avoid silly mistakes.

## SUBTRACTION EXERCISES

Answers appear on pages 195 and 196.

1. What is 623 – 211?

   **A.** 411

   **B.** 412

   **C.** 402

   **D.** 311

2. What is 843 – 109?

   **A.** 734

   **B.** 746

   **C.** 764

   **D.** 736

3. What is 752 – 221?

   **A.** 537

   **B.** 573

   **C.** 523

   **D.** 533

4. What is 807 – 142?

   **A.** 745

   **B.** 565

   **C.** 645

   **D.** 665

5. What is 900 – 65?

   **A.** 965

   **B.** 865

   **C.** 835

   **D.** 825

## MULTIPLYING BY MULTIPLES OF 10

Earlier in this book we discussed properties of multiplication and different techniques to memorize the basic multiplication facts. Now that we have a firm understanding of those basic facts it is time to extend them using multiples of 10. This is a fairly straightforward process so let's first start with an example:

$$
\begin{array}{r}
3\ 0 \\
\times\ \ \ \ 5 \\
\hline
\end{array}
$$

In the problem, we are multiplying $30 \times 5$. If you look closely at this problem you will recognize there is a basic multiplication fact hidden: $3 \times 5$. In this type of fact extension, we can ignore the 0 in 30 and solve the basic fact of $3 \times 5 = 15$. Now that we have the basic fact solved we can add the 0 back on to get a final answer of 150.

$$
\begin{array}{r}
3\ 0 \\
\times\ \ \ \ 5 \\
\hline
1\ 5\ 0
\end{array}
$$

This method of fact extension only works when the numbers involved end with zeros. You can ignore the ending zeros, solve the basic fact, then put the zeros back into the final answer.

There is also a standard form of multiplication that will work with all multiplication problems. You can utilize this method to solve fact extensions (shown above) or standard multi-digit problems. Let's take a look at another example using this method:

Step 1
Rewrite
vertically

$$
\begin{array}{r}
8\ 0 \\
\times\ \ \ \ 3 \\
\hline
\end{array}
$$

In the problem shown above, $80 \times 3$, we always solve by starting on the right side, just like in addition and subtraction. The key to solving using this method is that the 3 has to multiply by every digit on the top in the correct order. First we start with $3 \times 0$, which is 0. The 0 is placed directly under the ones column because we just solved in the ones column.

Step 2
Multiply by
ones position

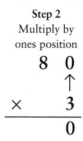

Next, the 3 must multiply by the 8. $3 \times 8 = 24$. Like in addition, the 4 comes down into the tens position (because we just solved the tens position) and the 2 is carried over to the hundreds position.

Step 3
Multiply by
tens position

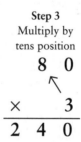

Finally, we get an answer of 240. This method works just as well as the fact extension method we first discussed; however, the benefit is that it can be used as a foundational procedure for solving more complex multiplication problems in the fourth grade. The fact extension method only works for problems where you have factors that include one digit (1–9) and all the rest are zeros, whereas this method works for all multiplication problems, regardless of the digits.

## MULTIPLICATION BY MULTIPLES OF 10 EXERCISES

Answers appear on pages 197 and 198.

1. What is 7 × 90?

   **A.** 63

   **B.** 630

   **C.** 540

   **D.** 720

2. What is 80 × 5?

   **A.** 405

   **B.** 400

   **C.** 40

   **D.** 320

3. What is 9 × 10?

   **A.** 900

   **B.** 9

   **C.** 90

   **D.** 0

4. What is 70 × 7?

   **A.** 560

   **B.** 450

   **C.** 400

   **D.** 490

5. What is 5 × 60?

   **A.** 300

   **B.** 12

   **C.** 3

   **D.** 30

# NUMBER AND OPERATIONS—FRACTIONS

Fractions are essential in our world. Many students consider their age in halves. ("I'm $8\frac{1}{2}$ years old.") Many, many recipes call for $\frac{1}{2}$, or $\frac{1}{4}$, or even $\frac{1}{8}$, of a full measure, be it flour or sugar or vanilla.

## DEFINITIONS: NUMERATOR, DENOMINATOR

A **Numerator** is the top value in a fraction. It refers to the number of items you are attempting to represent. In this way, the numerator is a multiplier.

A **Denominator** is the bottom value in a fraction. It refers to how many parts the whole item is broken into. In this way, the denominator is a divisor.

## UNDERSTANDING FRACTIONS

Fractions, like decimals, are often used to represent a number that is not a whole number. A fraction generally will represent only a portion, or part, of an entire object. For example:

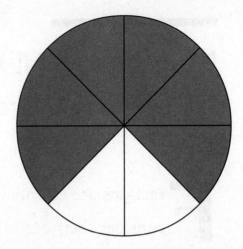

In the circle, you will notice that six pieces are shaded in. Let's pretend the circle is actually representing a pizza—and that the shaded pieces represent the amount of pizza remaining from the whole pizza. You also should realize that the pizza originally was eight slices in total, before anyone ate any of it. The pizza, when it arrived, was one whole pizza or $\frac{8}{8}$ (eight slices are there, out of a possibility of eight). Now that someone has eaten two slices (the white pieces), there are six remaining. So there are two possible fractions to represent this example:

**a.** The amount of pizza eaten so far is $\frac{2}{8}$ (or two out of a total of eight original pieces).

**b.** The amount of pizza remaining is $\frac{6}{8}$ (or six out of a total of eight original pieces).

Shown visually, here is what it would look like:

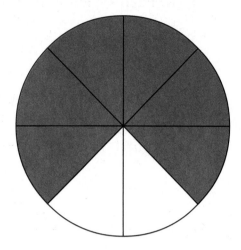

One whole pizza $= \dfrac{8}{8}$

Eaten pizza $= \dfrac{2}{8}$

Remaining pizza $= \dfrac{6}{8}$

## FRACTIONS ON A NUMBER LINE

Similarly, fractions may be used to represent a number line. A number line is a line that is divided into a certain number of sections. Some sections may be labeled with a value, while others are blank, so that you must determine and fill out the appropriate value. For example:

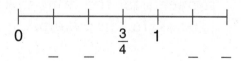

To solve a number line, it is necessary to locate two values that are next to each other. By doing this, we can determine how much each segment of the line is increasing or decreasing. In this example, $\dfrac{3}{4}$ and 1 are next to each other. The number 1 is the same thing as $\frac{4}{4}$.

Because $\frac{3}{4}$ and $\frac{4}{4}$ are next to each other, we now know

that the number line is increasing by $\frac{1}{4}$ in each step going

to the right (and oppositely decreasing by $\frac{1}{4}$ in each step

going to the left). To determine the first two missing

values, all we have to do is start at $\frac{3}{4}$ and subtract $\frac{1}{4}$

each time. The first value to the left of $\frac{3}{4}$ would be $\frac{2}{4}$

(also known as $\frac{1}{2}$), then to the left of that it would be $\frac{1}{4}$.
Now we can focus on the two remaining missing values
to the right of the 1. The first value to the right would be

$1\frac{1}{4}$; the next value to the right would be $1\frac{2}{4}$ (or

simplified $1\frac{1}{2}$).

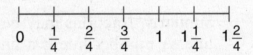

Similarly, if you are given a number line that has no
values whatsoever, and you are asked to fill in each of the
segments for the value it represents, it will be a similar
process. In this example,

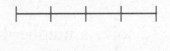

to determine the value of each section, you would count
up the total number of sections (*not* each line) and that

would be your denominator. The denominator is the value that represents the total number of segments in a given problem. If the line segment is broken into four segments, or sections, and no values are given, each segment is worth $\frac{1}{4}$. As you count to the right, adding each segment, the values increase.

## FRACTION EXERCISES

Answers appear on pages 198 and 199.

**1.** In the diagram,

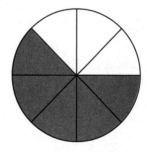

which value below would represent the fraction of the shaded region?

**A.** $\frac{1}{3}$

**B.** $\frac{3}{8}$

**C.** $\frac{5}{8}$

**D.** $\frac{8}{5}$

**2.** In the diagram,

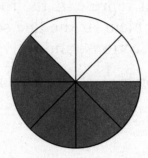

which value below would represent the fraction of the *un*shaded region?

A. $\dfrac{5}{8}$

B. $\dfrac{8}{3}$

C. $\dfrac{3}{8}$

D. $\dfrac{1}{8}$

**3.** In the number line shown,

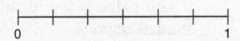

which fraction would represent the value of each segment?

A. $\dfrac{6}{1}$

B. $\dfrac{1}{1}$

C. $\dfrac{1}{3}$

D. $\dfrac{1}{6}$

**4.** In the number line shown,

what would be the value of the first two sections combined?

**A.** $\dfrac{1}{3}$

**B.** $\dfrac{3}{2}$

**C.** $\dfrac{2}{3}$

**D.** $\dfrac{3}{3}$

**5.** In the diagram shown,

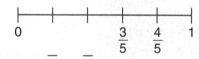

what is the value of the second missing space?

**A.** $\dfrac{1}{4}$

**B.** $\dfrac{1}{5}$

**C.** $\dfrac{2}{5}$

**D.** $\dfrac{5}{2}$

# EQUIVALENT FRACTIONS

Equivalent fractions are fractions that are equal in value even though they might look different. Regardless of the fact that the fractions look different, in terms of the values, they are actually representing the same overall value. Though this may sound confusing, it's actually quite simple. Let's take a look at an example by shading in $\frac{2}{4}$ and $\frac{1}{2}$ to get a visual representation. In the figures below, we can see what each fraction looks like because we are representing it on a circle. Remember, the denominator (bottom number) is the total number of pieces, while the numerator (top number) is the requested value (in this case, shaded).

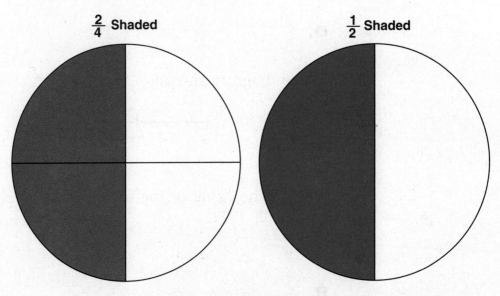

$\frac{2}{4}$ Shaded     $\frac{1}{2}$ Shaded

It is easy to see that these two fractions are equivalent because the shaded portion covers the exact same area in each. In this example, we have shaded these circles in to mirror each other. That way we can prove that the values are in fact equivalent.

We can also take this one step further and use a mathematical process to help determine whether two fractions are equivalent. This process is called cross-multiplication. In cross-multiplication, we multiply the denominator in fraction 1 by the numerator in fraction 2.

We then multiply the denominator in fraction 2 by the numerator in fraction 1. It's much easier than it sounds. Let's take a look at an example:

$$\overset{4}{\frac{2}{4}} \times \overset{4}{\frac{1}{2}}$$

By using cross-multiplication, we multiply 2 × 2 and get 4. We also multiply 4 × 1 and get 4. Because both answers are the same, these fractions are equivalent. If you are given a fraction and asked to name an equivalent fraction, you need to use the process of multiplication to calculate it. It is a two-step process. First, you must pick a whole number other than 0, then multiply the denominator and the numerator by that number. This will create a new fraction that is equivalent to the old one. For example, let's find a fraction that is equivalent to $\frac{2}{6}$. We'll randomly pick a number, say 3.

Note: the lower the number you pick, the easier the multiplication process will likely be. So we multiply the 2 × 3 to get 6 (the new numerator), and the 6 × 3 to get 18 (the new denominator). From this we can now say that $\frac{2}{6}$ is equivalent to $\frac{6}{18}$.

Now that we have a better understanding of fractions, we can take the next step of comparing them to determine greater-than, less-than, or equal (which can be represented >, <, =). This is the same process as described earlier to determine equivalent fractions. We are going to cross-multiply the fractions and then compare those two new answers. If they are the same, the fractions are equivalent. If one answer comes out higher, then the fraction on that side is larger. Let's look at some examples:

$$\frac{2}{3} \qquad \frac{1}{2}$$

If we cross-multiply these fractions, it would look like this:

$$\overset{4}{\frac{2}{3}} \diagup\!\!\!\diagdown \overset{3}{\frac{1}{2}}$$

Because 4 is larger than 3, that means the fraction on that side $\left(\dfrac{2}{3}\right)$ is larger. So $\dfrac{2}{3} > \dfrac{1}{2}$.

To prove the point, let's take a look at a visual representation of these two fractions compared side by side.

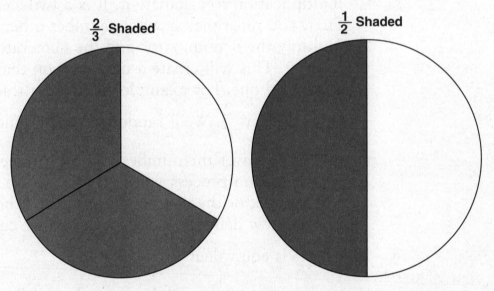

$\dfrac{2}{3}$ Shaded     $\dfrac{1}{2}$ Shaded

From the visual representation of these two fractions, we can see that the shaded region on the $\dfrac{2}{3}$ circle is larger than the shaded region on the $\dfrac{1}{2}$ circle. This further proves the mathematical process of cross-multiplication.

Here is another example:

$$\frac{7}{8} \qquad \frac{2}{2}$$

If we cross-multiply these fractions, it would look like this:

$$\overset{14}{\frac{7}{8}} \bowtie \overset{16}{\frac{2}{2}}$$

Since 16 is larger than 14, that means the fraction on that side $\left(\frac{2}{2}\right)$ is larger. So $\frac{7}{8} < \frac{2}{2}$.

Let's take a look at a visual representation of these two fractions compared side by side as further evidence.

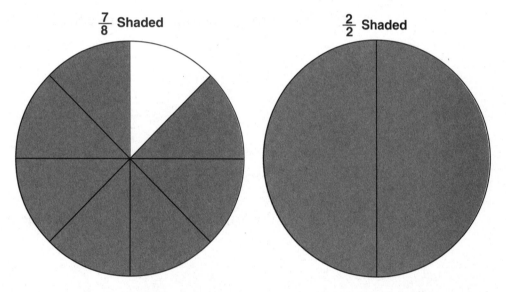

From the visual representation of these two fractions, we can see that the shaded region on the $\frac{2}{2}$ circle is larger than the shaded region on the $\frac{7}{8}$ circle.

## EQUIVALENT FRACTIONS EXERCISES

Answers appear on pages 200 to 202.

1. What symbol belongs in between these two fractions: $\frac{2}{3}$ —— $\frac{3}{4}$?

   A. >

   B. <

   C. =

   D. None of the above

2. What symbol belongs in between these two fractions: $\frac{3}{4}$ —— $\frac{1}{3}$?

   A. >

   B. <

   C. =

   D. None of the above

3. What symbol belongs in between these two fractions: $\frac{2}{2}$ —— $\frac{6}{6}$?

   A. >

   B. <

   C. =

   D. None of the above

**4.** What symbol belongs in between these two fractions: $\dfrac{4}{8}$ —— $\dfrac{3}{6}$?

    **A.** >

    **B.** <

    **C.** =

    **D.** None of the above

**5.** What fraction of this circle is shaded in?

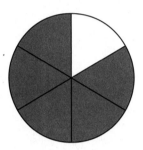

    **A.** $\dfrac{1}{5}$

    **B.** $\dfrac{1}{4}$

    **C.** $\dfrac{5}{6}$

    **D.** $\dfrac{5}{5}$

**6.** What fraction of this circle is *not* shaded in?

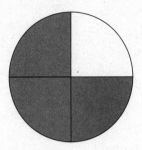

A. $\dfrac{1}{4}$

B. $\dfrac{3}{4}$

C. $\dfrac{1}{3}$

D. $\dfrac{3}{1}$

**7.** What fraction of this circle is shaded in?

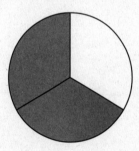

A. $\dfrac{2}{2}$

B. $\dfrac{1}{3}$

C. $\dfrac{2}{1}$

D. $\dfrac{2}{3}$

**8.** What fraction below is equivalent to $\frac{3}{4}$?

    **A.** $\frac{6}{8}$

    **B.** $\frac{4}{5}$

    **C.** $\frac{2}{3}$

    **D.** $\frac{1}{1}$

**9.** What fraction below is equivalent to $\frac{5}{8}$?

    **A.** $\frac{6}{9}$

    **B.** $\frac{4}{7}$

    **C.** $\frac{2}{4}$

    **D.** $\frac{15}{24}$

10. Which circle shows an unshaded area of $\frac{3}{8}$?

A.

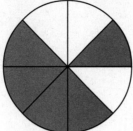

B.

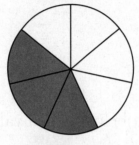

C.

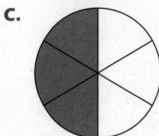

D.

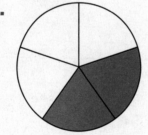

# MEASUREMENT AND DATA

## MEASUREMENT AND LENGTH

We frequently measure items in our lives. You could not buy a shirt, shoes, or a jacket without knowing your measurements. The temperature outside on any given day is a measurement that is on everyone's mind, to decide whether to wear that jacket or not.

When we measure length, there are two different systems we use: the metric system and the U.S. standard system.

## CUSTOMARY (U.S. STANDARD) SYSTEM

This is also called the English system because it originated in the United Kingdom. There are four main units of measure in this system:

1 inch (in.) is this length: _____

1 foot (ft) = 12 inches

1 yard (yd) = 3 feet or 36 inches

1 mile (mi) = 5,280 feet or 1,760 yards

Use your (customary) ruler to measure these paper clips:

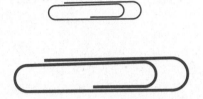

What did you find the length of the small paper clip to be? It should be one inch. Was the larger clip close to 2 inches, but not quite? The second clip should be $1\frac{7}{8}$ inches long.

Try this again, this time on two erasers:

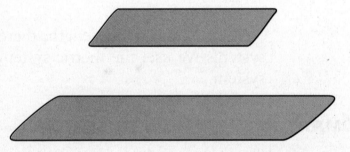

Did you measure the smaller eraser as 2 inches? Was the larger eraser $3\frac{1}{2}$? It should be that length.

## CUSTOMARY (U.S. STANDARD) SYSTEM EXERCISES

Answers appear on pages 202 to 204.

1. Measure these stamps with your (customary) ruler.
   Which one is between $1\frac{1}{2}$ and 2 inches wide?

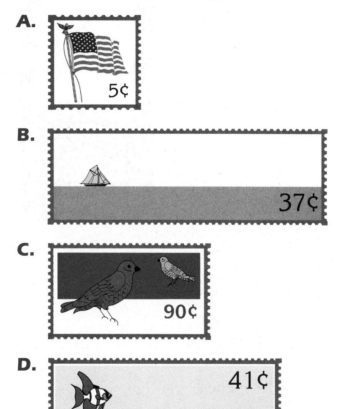

**A.** 5¢

**B.** 37¢

**C.** 90¢

**D.** 41¢

2. Here is an MP3 player. Using your (customary) ruler, tell how long it is.

A. 3 inches

B. $2\frac{1}{2}$ inches

C. $3\frac{1}{2}$ inches

D. 4 inches

3. What unit of measurement would we use to measure distance between cities?

A. Inches

B. Feet

C. Yards

D. Miles

**4.** Measure this ship with a (customary) ruler. Pick the pair of numbers that most closely matches the length and height of the ship.

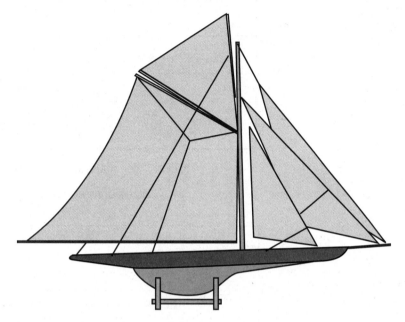

**A.** 2 inches, 4 inches

**B.** 3 inches, 5 inches

**C.** 4 inches, 3 inches

**D.** 5 inches, 2 inches

**5.** A notebook is one foot high. How many notebooks will make a yard?

**A.** 2

**B.** 3

**C.** 4

**D.** 5

**6.** Use the measurements below to prepare a line graph showing the sizes of each coin.

Coins are not drawn to scale.

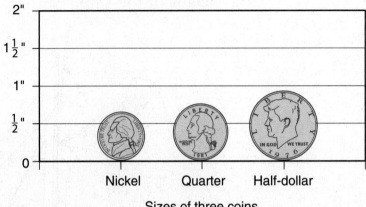

Sizes of three coins

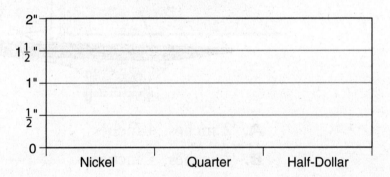

Sizes of three coins

**7.** Use the measurements below to prepare a line graph showing the sizes of the stamps.

Stamps not drawn to scale.

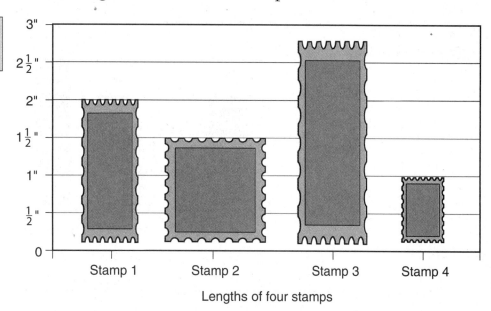

Lengths of four stamps

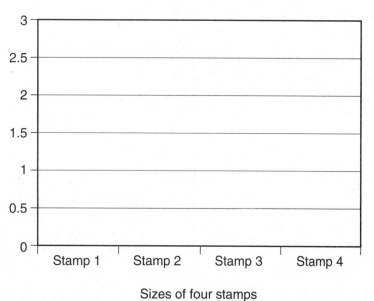

Sizes of four stamps

**8.** Use the measurements below to prepare a line graph showing the sizes of the erasers.

Erasers are not drawn to scale.

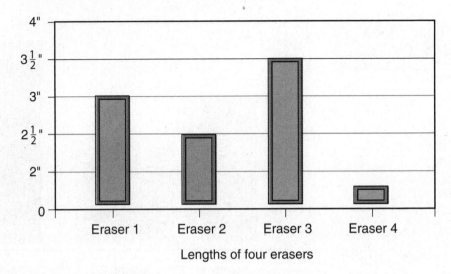

Lengths of four erasers

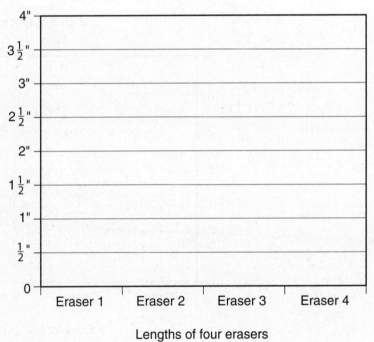

Lengths of four erasers

Coins are not drawn to scale.

**9.** Use the measurements below to prepare a line graph showing the sizes of the coins.

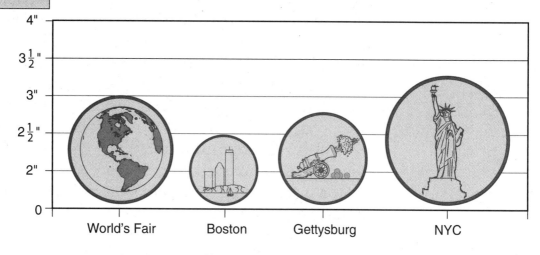

Sizes of four souvenir coins

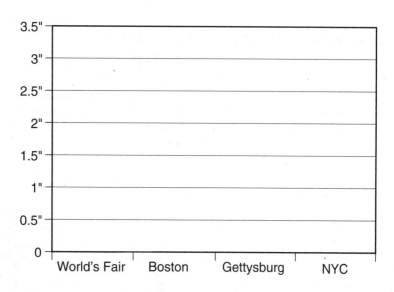

Sizes of four souvenir coins

## LIQUID VOLUME MEASURE (METRIC)

When we put water or any other liquid in a container (a cup, a bowl, or a pitcher, for instance) we measure the amount of liquid with a unit of measure called a **liter** (l). How big is a liter? It's a little more than a quart. It's actually one-tenth more than a quart (1 liter = 1.1 quarts). Most soft drinks and seltzer are sold in liter or 2-liter bottles. There are 1,000 milliliters in a liter. In other words, it takes 1,000 milliliters to make a liter. We measure small amounts of liquid (like a teaspoon of lime juice, for instance) in milliliters. A teaspoon of something is, in fact, 5 milliliters.

1,000 milliliters (ml) = 1 liter (l)

## LIQUID VOLUME MEASURE EXERCISES

Answers appear on pages 204 and 205.

1.

The pitcher above is 1 liter. How much liquid is in the pitcher? Give your answer in liters (l) and in milliliters (ml)

Write your answers here _____ (l) _____ (ml)

**2.**

The water bottle above is 1 liter. How much liquid is in the water bottle? Give your answer in liters (l) and in milliliters (ml).

Write your answers here _____ (l) _____ (ml)

**3.**

The pitchers above are each 1 liter. How much liquid is in the two pitchers combined? Give your answer in liters (l).

Write your answer here _____ (l)

**4.**

The three bottles above each are 1 liter. How much liquid is in the three bottles combined? Give your answer in liters (l).

Write your answer here _____ (l)

**5.**

The three jars above each are 1 liter. How much liquid is in the three jars combined? Give your answer in liters (l).

Write your answer here _____ (l)

**6.**

The three bottles above each are 1 liter. How much liquid is in the three bottles combined? Give your answer in liters (l).

Write your answer here _____ (l)

**7.**

The two pitchers above each are 1 liter. How much more liquid is in the first pitcher? Give your answer in liters (l) and in milliliters (ml).

Write your answers here _____ (l) _____ (ml)

8.

The three bottles above each are 1 liter. How much more liquid is in the first bottle than in the last two? Give your answer in liters (l) and in milliliters (ml).

Write your answers here _____ (l) _____ (ml)

9.

The two jars above each are 1 liter. If we were to multiply by 5, how many liters would we have?

Write your answers here _____ (l) _____ (ml)

**10.**

The bottle above is 1 liter. If we were to multiply by 6, how many liters would we have? Give your answer in liters (l).

Write your answer here _____ (l)

**11.**

The two pitchers above each are 1 liter filled with lemonade. If we are to provide drinks for eight children, and each child will receive an equal amount, how much lemonade will each child get? Give your answer in liters (l) and in milliliters (ml).

Write your answers here _____ (l) _____ (ml)

**12.**

The three pitchers above each are 1 liter filled with iced tea. If we are to provide drinks for 15 children, and each child will receive an equal amount, how much iced tea will each child get? Give your answer in liters (l) and in milliliters (ml).

Write your answers here _____ (l) _____ (ml)

---

### TIP

**Estimation of volume capacities**—It is useful in various situations to estimate what size bowl or pitcher is needed to hold an amount of liquid.

## VOLUME ESTIMATION EXERCISES

Answers appear on pages 205 and 206.

1. Which of the items above would you use to measure 5 teaspoons of a liquid?

Write your answer here _____

1 liter bottle

2. Which of the items above would you use to measure 2 liters of a liquid?

Write your answer here _____

**3.** Which of the items above would you use to fill up a bathtub?

Write your answer here _____

1 liter bottle

**4.** Which of the items above would you use on a hike to carry water to drink on a trail?

Write your answer here _____

Choose the correct liquid measurement for each of the following:

**5.** 3 l or 15 ml?

Write your answer here _____

1 liter bottle

**6.** 50 ml or 7 l?

Write your answer here _____

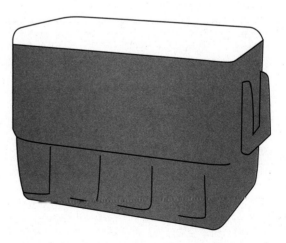

**7.** 6 l or 50 ml?

Write your answer here _____

**8.** 3 l or 60 ml?

Write your answer here _____

## REASONING

Answers appear on page 206.

**9.** Which unit would you use if you were filling up a bucket: milliliters or liters? Explain your answer:

Write your answer here _____

_____

_____

**10.** Which unit would you use if you were filling up a small cup: milliliters or liters? Explain your answer:

Write your answer here _____

_____

_____

## MASS MEASURE (METRIC)

We measure mass, the amount of matter in an object, using a unit called the **gram** (g). The gram is quite small. It takes 28 grams to equal one ounce (in the customary system). A **kilogram** (kg) is 1,000 grams, and it is used in many instances for mass. A kilogram is about 2.2 pounds (in the customary system).

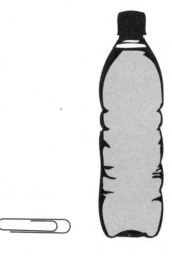

1 liter bottle

A large paper clip has a mass of about a gram (g).

A liter of water has a mass of a kilogram (kg).

## MASS ESTIMATION EXERCISES

Answers appear on page 206.
Give the better estimate for each of the following:

**1.**

5 g or 550 kg? Answer _____

**2.**

20 g or 2 kg? Answer _____

**3.**

4 kg or 150 g? Answer _____

**4.**

80 g or 500 g? Answer _____

**5.**

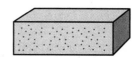

140 g or 2 kg? Answer _____

**6.**

1,000 kg or 50 kg? Answer _____

**7.**

140 g or 50 g? Answer _____

**8.**

450 g or 3 kg? Answer _____

## MASS CALCULATION EXERCISES

Answers appear on pages 206 and 207.

1. Filomina had 300 g of sugar in her sugar bowl. She bought 450 g of sugar. How much sugar did she have after she combined the two amounts?

    Answer _____

2. Tenille had 20 kg of potatoes for a picnic she was planning, but she realized she needed more. She went out and bought another 12 kg. How many kg of potatoes did she have then?

    Answer _____

3. Jorge used up 120 g of coffee for a small get-together with his friends. If he originally had 450 g of coffee in his coffee tin, how much does he have now?

    Answer _____

4. A board Edwina was cutting was 3 kg. She cut some of it off, and then the board was 1.8 kg. How much was the piece she cut off?

    Answer _____

5. Jessica is a plumber. She needs to cut a piece of pipe to fit into a bathroom fixture. The piece of pipe she needs to cut measures 18 kg. She decides to cut off 9 kg. What is the mass of the pipe now?

    Answer _____

**6.** Danielle is having a party with her friends. She has measured out rice for all the guests. She wants 56 g of rice for each friend, and she has six friends coming. If she makes the same portion for herself, how much rice will she measure out?

Answer _____

**7.** Coming in from a cold afternoon of sledding, Walter measured out 27 g of hot cocoa mix for each of his eight friends and himself. How many grams of hot cocoa did he measure out?

Answer _____

**8.** Tim has 30 g of tea in his tea caddy. He wants to make tea for himself and his seven friends. Each cup requires 1.5 g of tea. How much tea will he use? Will there be any tea left in the caddy? How many g will be left?

Answer _____

**9.** Cosimo went to the store to get some salad for a get-together with his family. He got 78 g of salad. Each member of his family had an equal share of the salad. If the number in his family equals six (including Cosimo), how many g of salad did each member get?

Answer _____

**10.** The bricks that Harry was working with were just a little too big for the space he needed to place them into. Each brick was 8″ long, but he was able to cut a small slice so it fit into the space. The bricks were 2.5 kg each, but when he cut them to the right size they were 1.9 kg. How much mass did he cut off?

Answer _____

## TIME

You use time every day. You get up at a certain time to go to school for a certain number of hours, and usually eat at certain times. **Time** is measured in hours, minutes, and seconds, but we won't be looking at seconds. It's also measured in days, weeks, months, and years, but we will not be talking about those larger units in this section. Hours are measured from midnight with the designation "A.M."; hours from noon have the designation "P.M." Hours are counted 12 to 1, 2, and so forth up to 11. Minutes are designated 0 to 60.

As an example of time measurement, let's look at a digital clock.

In this clock, the first number (7) gives the hour. The next two numbers (43) give the minutes. The designation P.M. means that it is afternoon or evening. So, it is 7 hours and 43 minutes after noon.

Another example looks at a time before noon.

In this clock, the first number (8) gives the hour, and the next two numbers (23) give the minutes. A.M. means it is before noon, but it really tells us it is 8 hours and 23 minutes *after midnight*. To find out how far it is before noon, you would need to subtract it from 12:00. Remember, though, that hours go only up to 12, and minutes and seconds go only up to 60. So, 11 − 8 = 3, and 60 − 23 = 37. This clock tells us that we are 3 hours and 37 minutes before noon.

## TIME EXERCISES

Answers appear on page 207.

1. Matthew woke up and looked at his clock. It read 7:34 A.M. He took a shower, dressed, and went down to make breakfast. When he looked at the clock again, it read 8:46 A.M. How much time had gone by since Matthew had gotten up?

   **A.** 1 hour, 15 minutes

   **B.** 2 hours, 20 minutes

   **C.** 1 hour, 12 minutes

   **D.** 1 hour, 10 minutes

2. Darlene was working on her homework. She looked at the clock, and it read 8:27 P.M. After studying and writing out some homework sheets, she looked at the clock again, and it read 9:38 P.M. How much time had Diana worked on her homework?

   **A.** 1 hour, 35 minutes

   **B.** 1 hour, 11 minutes

   **C.** 1 hour, 27 minutes

   **D.** 1 hour, 12 minutes

3. Ted coaches wrestling and was giving his wrestlers some drills. He started them at 3:15 P.M. and drilled them for 23 minutes. What time was it then?

   **A.** 3:23 P.M.

   **B.** 3:33 P.M.

   **C.** 3:48 P.M.

   **D.** 3:38 P.M.

4. Dan was building a bridge along with seven other boys. They worked together $3\frac{1}{2}$ hours, and they finished the bridge. If you were to spread the hours out, how many hours would that be?

   **A.** 28 hours

   **B.** $21\frac{1}{2}$ hours

   **C.** 23 hours

   **D.** 20 hours

**5.** Joe was studying for his math exam. He started at 7:15 P.M., and studied 2 hours and 35 minutes. What time was it then?

**A.** 9:45 P.M.

**B.** 9:40 P.M.

**C.** 9:50 P.M.

**D.** 9:55 P.M.

Time is also measured on a clock (sometimes called an analog clock).

In all clocks of this kind, the small hand indicates the hour, and the large hand indicates the minutes. In the clock above, the small hand points to the 3, and the large hand points to the 12, meaning that it is 3 o'clock exactly (no minutes past 3).

In this clock, the small hand points a little past 3, and the large hand points to 1. All the numbers 1 through 12 indicate 5 minutes, so this clock indicates 5 minutes past 3.

In this clock, the small hand is about a quarter past the 5, and the large hand is pointing to the 3. The large hand indicates 15 minutes past the hour of 5. So this clock indicates a time of 5:15.

In this clock, the small hand is more than half past the 4, and the large hand is pointing to the 7. The large hand indicates 36 minutes past the hour of 4. So this clock indicates a time of 4:36.

In this clock, the small hand is long past the 6, almost pointing to 7. This means that it is still on the hour of 7. The large hand is in between the 10 and the 11. The large hand indicates 53 minutes past the hour of 6. So this clock indicates a time of 6:53.

## EXERCISES FOR TIME (ANALOG CLOCK)

Answers appear on page 208.

1.

Write the time showing on the above clock here

_____

2.

Write the time showing on the above clock here

_____

3.

Fred saw the clock above. What time did he read?

_____

4.

Yanni saw the clock above. What time did he read?

_____

**5.**

Pat finished lunch, and looked at the clock above. What time did she read? _____

**6.**

Tamara finished her shopping and noticed the clock above. What time did she see? _____

**7.**

From his desk, Mahmoud glanced at the clock above. What time did he read? _____

**8.**

Jamie finished his breakfast and glanced at the clock above. What time did he observe? _____

**9.**

Francois finished his studying and looked at the clock above. What time did he read? _____

**10.**

Helga finished class and noticed the time. What time did she see? _____

## EXTENDED CONSTRUCTED RESPONSE (ECR) QUESTIONS

ECR questions require you to give an answer and then to explain how you got the answer. You could explain your answer using a chart or graph, using pictures, or using words. The examiners of the test award points for the extended constructed response questions on a scale of 0 to 3. They award more points for the more complete explanation. They can award only 1 point to a student who gives the correct answer with no explanation.

## SAMPLE EXTENDED CONSTRUCTED RESPONSE QUESTIONS (FOR MEASUREMENT AND DATA)

1. Harry is painting the floor of a closet in his house. The dimensions of the floor are shown below.

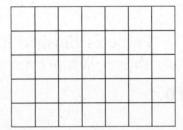

Each square represents a square foot. How would you find out the area of this floor?

*Answer to question 1.* You could count the number of squares: 35. Another way to solve this would be to multiply the number of feet in the length, which is 7, by the number of feet in the width, which is 5. 7 × 5 = 35. A third way might be to add columns of five, seven times, or to add rows of seven, five times. This third way is really multiplication, of course, but it sometimes is used by students who are not completely comfortable multiplying.

2. Look at the figures below.

 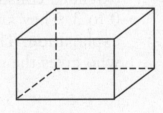

Name the two figures.
Tell how many edges each figure has.
Tell how many faces each figure has.
Tell one way each figure is the same.
Tell one way each figure is different.

*Answer to question 2.* There are five questions here. The answers to the first question are: The first figure is a three-sided pyramid, and the second figure is a rectangular prism.

The second question asks how many edges each figure has. The pyramid has six edges, and the rectangular prism has 12 edges.

The third question asks how many faces each figure has. The pyramid has four faces, and the rectangular prism has six faces.

The fourth question asks one way in which the figures are the same. There are several ways that the figures are the same:

**A.** Both figures have only straight edges (no curves).

**B.** Both figures have only flat faces (again, no curving lines).

**C.** Both figures are three-dimensional.

The fifth question asks how the two figures are different. There are several ways in which the figures are different:

**A.** The pyramid is smaller than the rectangular prism.

**B.** The pyramid has a smaller number of edges than the rectangular prism.

**C.** The pyramid has a smaller number of faces than the rectangular prism.

**D.** The pyramid is made of triangles, whereas the rectangular prism is made of rectangles and squares.

**E.** For the pyramid, the angles made by the edges are all acute angles, whereas for the rectangular prism the angles made by the edges are all right angles.

**3.** Effram visited a house of worship in a small village. The building is pictured below.

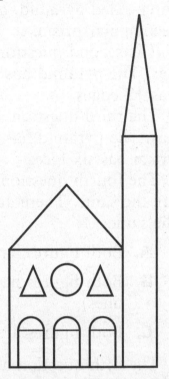

In this house of worship, identify all the kinds of geometric figures you can.

Tell how many of each kind of figure there is.

Tell which figures are congruent to each other.

*Answer to question 3:* There are three questions here. The answer to the first question is that there are squares, rectangles, triangles, circles, and half-circles in this picture. The answer to the second question is that there is one square, four rectangles, one circle, four triangles, and three half-circles in the picture. To answer the third question, we note that the three doors are congruent rectangles. Likewise, the half-circle windows above each door are congruent to each other. Finally, the two triangle windows on either side of the circular window are congruent to each other.

**4.** Ted is planning on building a deck in the back of his house. The deck will be 54 feet around the edge, and will be 15 feet long. Tell how wide the deck will be. Ted wants to put little lights every 3 feet around the edge of the deck. How many will he need? How did you get your answer?

*Answer to question 4.* Since the perimeter is 54 feet and the length is 15 feet, the two sides of the length are $2 \times 15 = 30$ feet, and $54 - 30 = 24$ so the two sides of the width are 24 feet and $24 \div 2 = 12$, so the width is 12 feet.

To determine the number of lights around the edge, note that, along the length, there will be a light at the corner, and one every 3 feet, over to the second corner. Marking those lights, they are at 0, 3, 6, 9, 12, and 15 feet. So that makes six lights on each length. Since there are two lengths, there are 12 lights needed for the lengths. Now we need to place lights along the width. We already have lights at each corner, so we only need to place lights at the feet markers of 3, 6, and 9 (because at 0 and 12 there are already lights). That makes three lights along each width. There are two widths, so we need a total of six more lights for the two widths. So the total number of lights we need along the edge of the deck is $12 + 6 = 18$ lights.

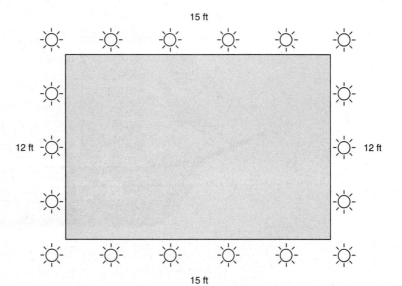

**5.** How many cubes are needed to make this figure?

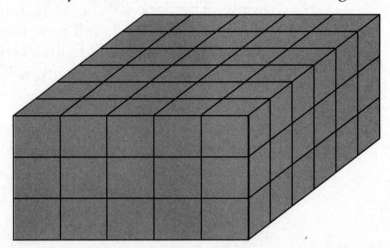

*Answer to number 5.* Note that the prism is six cubes long and six cubes wide. So each layer is 6 × 5 = 30 cubes. Now, the prism is three cubes high, so we need to multiply 30 by 3: 30 × 3 = 90. The prism requires 90 cubes.

## EXTENDED CONSTRUCTED RESPONSE QUESTION (MEASUREMENT AND DATA) EXERCISES

Answers appear on page 210.

**1.** Look at the ship below.

List and name all the geometric figures.
Tell which figures are equivalent to others.
Explain why they are equivalent.

**2.** Consider the two figures below.

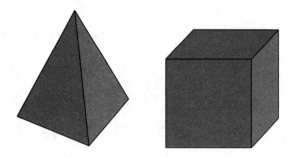

Tell how many edges each figure has.
Tell how many faces each figure has.
Tell one way each figure is the same.
Tell one way each figure is different.

## DATA ANALYSIS

We use data analysis all the time to order our world.
Sometimes teachers arrange students in their classes in
alphabetical order, and sometimes they order them by
height. At the ice cream shop, you can find out how
many possibilities you have with the ice cream flavors
available, the cone types, and the toppings, using discrete
mathematics. In the remainder of this chapter, we'll be
looking at these topics.

When you collect information, such as the heights of all
your classmates, or how many have red shirts and how
many have blue shirts, you are collecting **data**. When you
put that data in a chart or a graph of some kind and then
make judgments based on the graph, that is **data analysis**.
Let us look at a simple example of this.

In a class of 22 students, it was found that:

14 students have brown eyes,
six students have blue eyes, and
two students have green eyes.

To make these data easier to understand (or see), it can be put in the form of a pictograph:

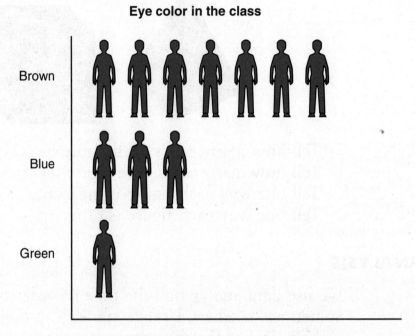

**Eye color in the class**

One student stands for two students

Pictographs often will have pictures stand for the data they represent. The pictures often will be a picture of the thing they represent. So, in this pictograph, there are pictures of students, and each picture of a student stands for two students. In national pictographs, one person might stand for a thousand (or even a million) people.

These data can also be put in the form of a table:

| Brown | 14 |
|-------|----|
| Blue  | 6  |
| Green | 2  |

The data can also be in the form of a bar graph:

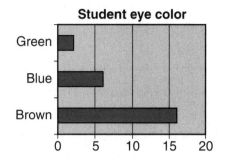

Now we could go the other way. We could look at a graph and make observations based on that graph. Consider this graph.

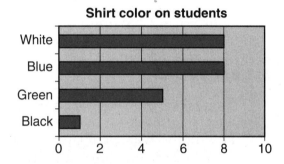

In this example, how many students would you say:

Are wearing black shirts?
Are wearing green shirts?
Are wearing blue shirts?
Are wearing white shirts?

There is one student wearing a black shirt.
There are five students wearing green shirts.
There are eight students wearing blue shirts.
There are eight students wearing white shirts.

We can make certain calculations based on these data as well. For example:

How many more students are wearing blue than black?

Since there are eight students who are wearing blue and one student who is wearing black, there are 8 − 1 = 7. Seven more students are wearing blue than are wearing black.

How many students are in the class?

Since there are eight students wearing white, eight students wearing blue, five students wearing green, and one student wearing black, then we have

$$8 + 8 + 5 + 1 = 22$$

There are 22 students in the class.

## DATA ANALYSIS EXERCISES

Answers appear on pages 209 and 210.

1. Emelio was ordering pizza for the class party, and wrote down what all the students wanted. He put the information in the following table:

| Plain | X X X X X |
|---|---|
| Pepperoni | X X X X X X X X X |
| Mushrooms | X X X |
| Sausage | X X X X X X X |

Based on this information, which of the following is true?

**A.** There are more students who like sausage than pepperoni.

**B.** There are fewer students who like mushrooms than plain.

**C.** There are more students who like plain than sausage.

**D.** There are more students who like mushroom than sausage.

**2.** Kara drew up a chart that represents the number of students born in the different months of the year:

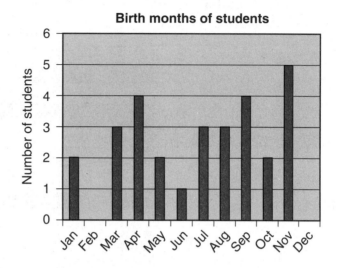

How many students were born in the first half of the year (January to June)?

**A.** 12

**B.** 11

**C.** 14

**D.** 15

**3.** Warren is a traffic officer and was looking at the cars driving down Gravel Rd. He wrote his findings in a chart:

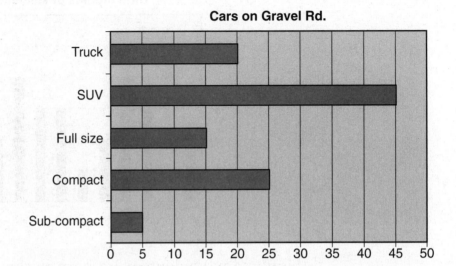

Which statement is true, based on the data?

**A.** Trucks are the most popular vehicles on Gravel Rd.

**B.** Compacts are the least popular vehicles on Gravel Rd.

**C.** Full-size cars are the least popular vehicles on Gravel Rd.

**D.** The SUVs are the most popular vehicles on Gravel Rd.

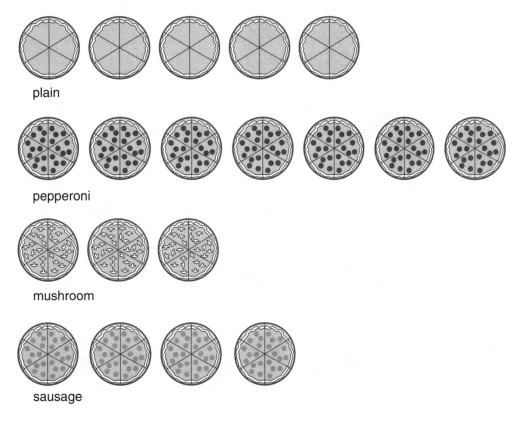

One pizza stands for four pizzas

4. The above figure represents the pizzas ordered by the Walker St. School for their Field Day. As the legend says, each pizza stands for four pizzas. Using this pictograph, please answer the following questions:

   **A.** How many plain pizzas did the Walker St. School order? _____

   **B.** How many sausage pizzas did the Walker St. School order? _____

   **C.** How many <u>more</u> plain pizzas than mushroom pizzas did the Walker St. School order? _____

   **D.** How many <u>more</u> pepperoni pizzas than sausage pizzas did the Walker St. School order? _____

   **E.** How many pizzas altogether did the Walker St. School order? _____

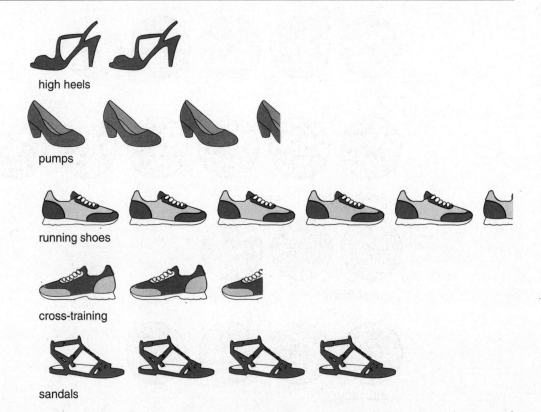

high heels

pumps

running shoes

cross-training

sandals

One shoe stands for 12 pairs of shoes

**5.** The above figure represents the different amounts and types of shoes in the ladies shoe department of Sibley's department store.

**A.** How many pairs of running shoes are in the ladies shoe department? _____

**B.** How many pairs of running shoes and cross-training shoes combined are there? _____

**C.** How many <u>more</u> pairs of pumps than high heeled shoes are in the ladies shoe department? _____

**D.** How many <u>fewer</u> pairs of sandals than running shoes are there? _____

**E.** If we consider high heeled and pump shoes to be dress shoes, and running shoes, cross-training shoes, and sandals to be casual shoes, how many <u>fewer</u> pairs of dress shoes than casual shoes are in the ladies shoe department? _____

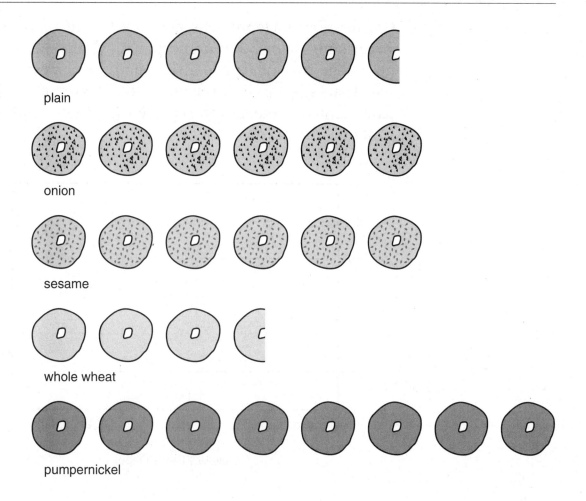

Bengal's Bagels. Each bagel represents ten bagels

6. The pictograph above represents the number of bagels sold by Bengal's Bagels on one cold November day.

   **A.** How many onion bagels did Bengal's sell? _____

   **B.** How many <u>more</u> pumpernickel than plain bagels were sold? _____

   **C.** How many more onion than whole wheat bagels did Bengal's sell? _____

   **D.** How many more onion and sesame bagels than plain bagels did Bengal's sell? _____

   **E.** What was the total number of bagels sold by Bengal's that cold November day? _____

**7.** Giselle's Pants Depot sells just three kinds of pants: sweatpants, jeans, and cargo pants. When she opened up on the first of April, Giselle had 30 pairs of sweatpants, 60 pairs of jeans, and 50 pairs of cargo pants. Draw a bar graph representing these amounts of pants on the graph below:

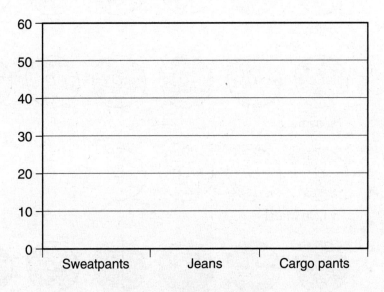

Giselle's Pants Depot

**8.** Elaine decided to clean up and organize her books. She had 20 cookbooks, 15 novels, 10 travel books, and 35 nursing books. Draw a bar graph representing these books.

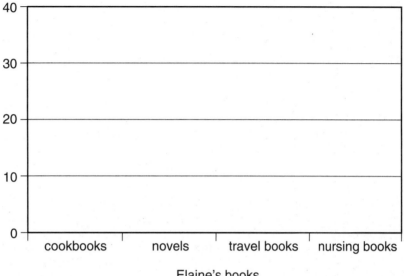

Elaine's books

# GEOMETRY

Geometric shapes are everywhere in our world. We play with lots of different sized balls, which are spheres. Sugar cubes are just that: cubes. We write on pieces of paper, which most of the time are shaped like rectangles. We eat pie, which is shaped like a circle. Along the road, we see traffic signs in many shapes (triangle, square, octagon). In this chapter, we will explore some characteristics of geometry.

## DEFINITIONS: POLYGONS, VERTICES, AND OTHER GEOMETRIC TERMS

A **polygon** is a many sided two-dimensional figure. In other words, it is a closed figure of three or more straight lines on a flat piece of paper. Examples are a triangle, a pentagon, and an octagon.

An **edge** is a side of the polygon. Edges is the plural of "edge." An edge is a line segment, that is, a small piece of a straight line.

A **vertex** is where two edges meet. Vertices is the plural of "vertex." A vertex forms an angle on the polygon.

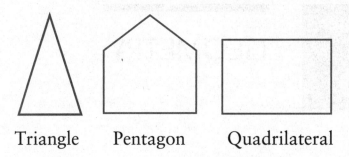

Triangle    Pentagon    Quadrilateral

## POLYGONS

There are many polygons. The following table names just a few and tells how many edges and vertices each one has.

| Polygon | Edges | Vertices |
|---|---|---|
| Triangle | Three | Three |
| Quadrilateral | Four | Four |
| Pentagon | Five | Five |
| Hexagon | Six | Six |
| Octagon | Eight | Eight |
| Decagon | Ten | Ten |

When we deal with polygons, the word "regular" is important. A **regular polygon** is a polygon that has all edges the same size, and the angles at each edge are the same size. A few examples of these are:

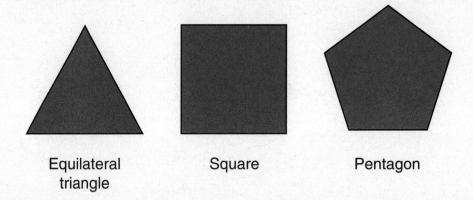

Equilateral    Square    Pentagon
triangle

## QUADRILATERALS

Quadrilaterals are, as just defined, four-sided polygons, but there are five special types of quadrilaterals that should be explained.

A **trapezoid** is a quadrilateral that has one pair of opposite legs that are parallel.

A **parallelogram** is a quadrilateral that has both pairs of opposite legs parallel.

A **rhombus** is a quadrilateral that is a parallelogram, but in addition, all four legs are the same length.

A **rectangle** is a quadrilateral that is a parallelogram, but in addition, all four angles are right angles (90°).

A **square** is a quadrilateral that combines one property of a rhombus with another of a rectangle. A square has all four sides that are equal in length, as well as four right angles.

## CIRCLES

Since they have no straight sides and no vertices, **circles** have special names for their parts. The edge of the circle is one continuous curved line that goes all the way around the circle. The **radius** of a circle is the distance from the center of the circle to the edge.

The **diameter** of a circle is the distance from one edge of a circle, through the center, and over to the other edge. The diameter is, therefore, two times the size of the radius.

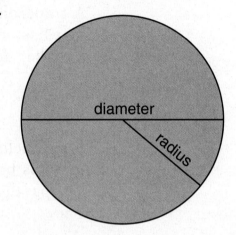

## POLYGON EXERCISES

Answers appear on page 211.

**1.** Which of the following is a polygon?

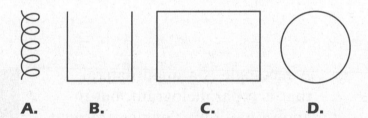

A.    B.    C.    D.

**2.** If you put two equilateral triangles together, what shape would you form?

   **A.** A square

   **B.** A rectangle

   **C.** A hexagon

   **D.** A rhombus

**3.** What polygon has six sides?

    **A.** A hexagon

    **B.** An octagon

    **C.** A pentagon

    **D.** A quadrilateral

**4.** Which shape is *not* a polygon? Circle your answer.

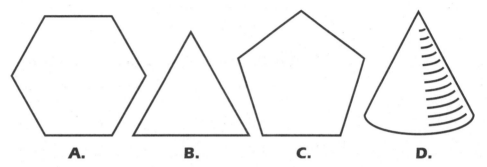

    **A.**       **B.**       **C.**       **D.**

**5.** Tell how many sides each polygon has. Name each polygon and write it on the line.

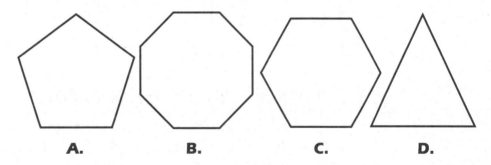

    **A.**       **B.**       **C.**       **D.**

_____    _____    _____    _____

## QUADRILATERAL EXERCISES

In the following exercises, circle your answer.
Answers appear on page 211.

**6.** Which of these shapes is a rhombus?

**7.** Which of these shapes is a rectangle?

**8.** Which of these shapes is a trapezoid?

**9.** Which of these shapes is a square?

**10.** Which of these shapes is a parallelogram?

# COORDINATE GRIDS

The **coordinate grid** is the *x-y* plane, and points on that grid can be located in terms of two points, the *x* point and the *y* point. The *x*-axis runs horizontally, and the *y*-axis runs vertically. When referring to a location on the grid, the *x* point is named first, followed by the *y* point. The resulting location is called an **ordered pair**. An example of this is:

Three points are illustrated here:

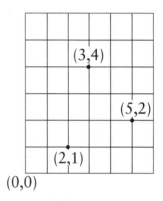

Point (3, 4)
Point (5, 2)
Point (2, 1)

# LOCATING POINTS ON THE COORDINATE GRID EXERCISES

Locate these five points on the coordinate grid.

Answers appear on page 212.

**1.** (2, 3)

**2.** (5, 1)

**3.** (4, 4)

**4.** (3, 1)

**5.** (6, 0)

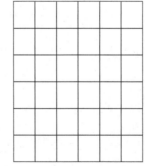

## PERIMETER OF A FIGURE

The **perimeter** of a figure is the total distance around an object. Consider this figure:

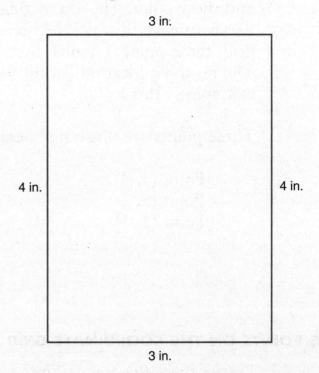

3 in.

4 in.

4 in.

3 in.

You see here that the perimeter of this rectangle is $3 + 3 + 4 + 4 = 14$ inches around. Another way to look at this is to add one side to another and then double it. So $(3 + 4) \times 2 = 14$ inches.

Let's look at another example:

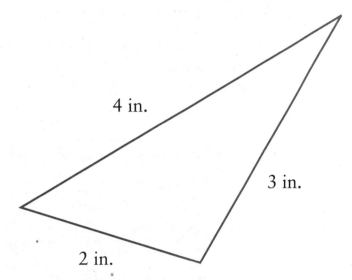

You see here that the perimeter of this triangle is
2 + 3 + 4 = 9 inches around.

Curved figures have a perimeter, as well.

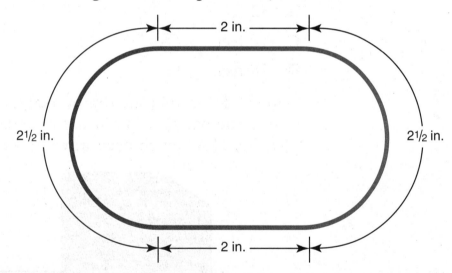

You see here that the top and bottom are 2 in., and
$2\frac{1}{2}$ in. on each end, so that the perimeter is

$$2 + 2\frac{1}{2} + 2 + 2\frac{1}{2} = 9 \text{ in. around.}$$

The perimeter of a circle can be estimated by multiplying
the diameter by 3.

## PERIMETER OF SHAPES EXERCISES

Answers appear on page 212.

1. Sean is putting in a flower bed and wants to put a fence around it. The flower bed is shown below. How many feet of fence should Sean buy?

13 feet

6 feet

A. 38 feet

B. 32 feet

C. 19 feet

D. 78 feet

2. Gilda is decorating the top of a square box, shown below. She wants to put decorative paper around the box top. How much decorative paper does she need?

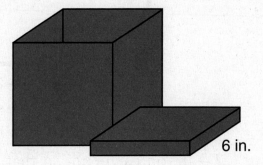

6 in.

A. 36 inches

B. 30 inches

C. 28 inches

D. 24 inches

**3.** Leroy runs cross country. To practice, he runs around the school track, pictured below. How far around is this track?

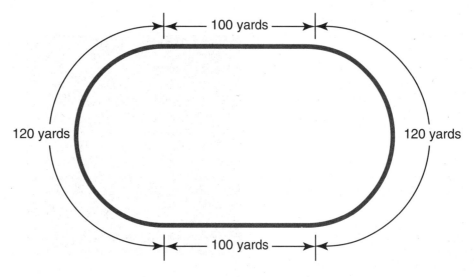

**A.** 220 yards

**B.** 340 yards

**C.** 440 yards

**D.** 400 yards

**4.** Andrea bought a pair of shoes. The rectangular box the shoes came in is pictured. What is the perimeter of the box?

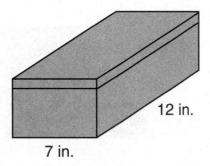

**A.** 40 inches

**B.** 38 inches

**C.** 34 inches

**D.** 19 inches

**5.** George is putting a frame around his favorite poster. The poster is shown below. How much framing material will George be getting?

20 inches

20 inches

**A.** 400 inches

**B.** 140 inches

**C.** 120 inches

**D.** 80 inches

**6.** Greg has odd-shaped tables in his restaurant. The tables are shaped like a trapezoid. What is the perimeter of the tables?

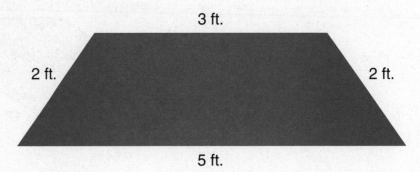

3 ft.

2 ft.          2 ft.

5 ft.

**A.** 6 feet

**B.** 10 feet

**C.** 12 feet

**D.** 18 feet

**7.** Joe sells stained glass windows. The one he has on display is a regular hexagon. What is the perimeter of the window?

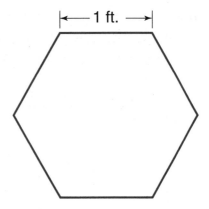

**A.** 8 feet

**B.** 6 feet

**C.** 4 feet

**D.** 10 feet

## DETERMINE THE AREA OF SHAPES ON A SQUARE GRID

**Area** is the measure of a two-dimensional shape. It is measured in square units. This is different from the linear measure in earlier sections. To measure area, simply multiply the **length** by the **width**. Some figures, like squares and rectangles, are easy to measure, thus:

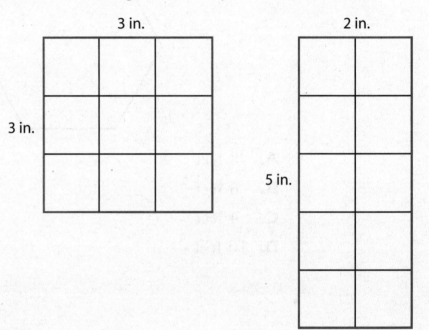

You see that the square is 3 inches on each side; you could count the squares inside to get 9. There is also an equation for the area of the square:

$$A = 3 \times 3 = 9 \text{ square inches},$$
which is also written as 9 in.$^2$

In similar fashion, the rectangle is 2 inches on one side, 5 inches on the other side, so counting the squares gives you 10. Again, there is an equation for the area of the rectangle:

$$A = 2 \times 5 = 10 \text{ square inches},$$
which is also written as 10 in.$^2$

You can see, therefore, that square measure is a *different kind* of measure than linear measure. Square measurements always use square units, whether it be square inches (in.²), square feet (ft²), or square centimeters (cm²).

Some figures, however, are not as easy to measure. Since area is a measure of the length times the width, how are we to measure a circle, where there are only curved lines?

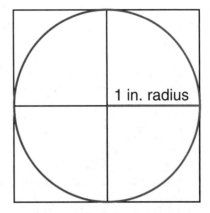

As before, one way of doing this is to draw the figure on a coordinate grid. That way we can count the number of squares inside the figure, and add partial squares to get a reasonable estimate of the area.

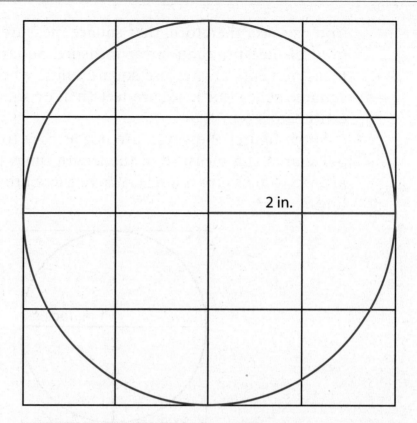

Here is another figure, which is not so difficult to find the area of:

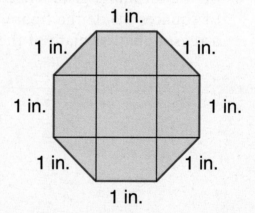

As you can see, with an octagon, it is easy to line up the squares from the grid and get a more accurate measure of the area.

## AREA OF SHAPES ON A SQUARE GRID EXERCISES

Answers appear on pages 213 to 215.

> **Note:** When doing these exercises, make sure you write down which units the problem is displaying. Always use the correct units that the exercises require.

1. The CD case above has a grid laid over it. Each square is 1 square inch. Calculate the area of the CD in square inches.

_____

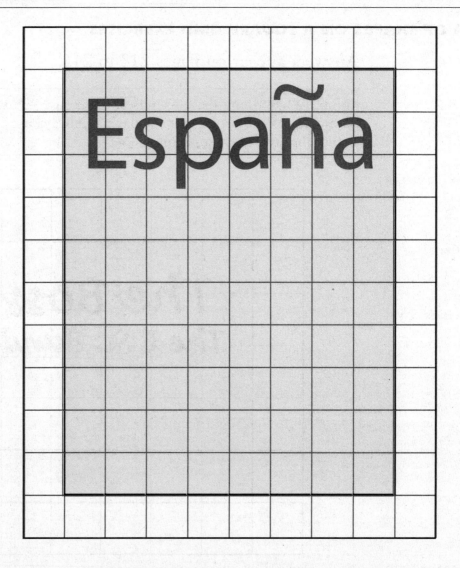

**2.** The book above has a grid laid over it. Each square is 1 square inch. Calculate the area of the book in square inches.

_____

**3.** The same book is _____ inches wide and _____ inches high. Multiply these two, and place your answer here _____. Is it the same as the area you calculated in question 2?

**4.** The cereal box above has a grid laid over it. Each square is 1 square inch. Calculate the area of the cereal box in square inches.

_____

**5.** The cereal box is _____ inches wide and _____ inches high. Multiply these two, and place your answer here _____. Is it the same area as you calculated in question 4?

**6.** The macaroni and cheese box above has a grid laid over it. Each square is 1 square centimeter. Calculate the area of the box in square centimeters.

_____

**7.** The macaroni and cheese box is _____ cm wide and _____ cm high. Multiply these two, and place your answer here _____. Is it the same area as you calculated in question 6?

**8.** The flyer above has a grid laid over it. Each square is 1 square inch. Calculate the area of the flyer in square inches.

_____

**9.** The flyer is _____ inches wide and _____ inches high. Multiply these two, and place your answer here _____. Is it the same area as you calculated in question 8?

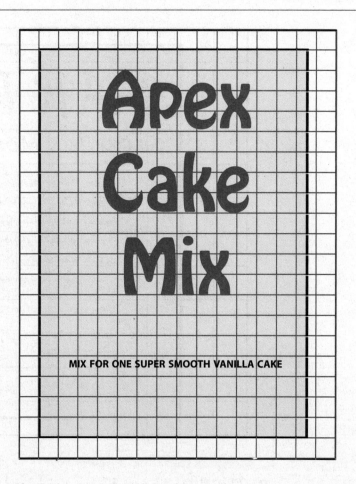

**10.** The cake box above has a grid laid over it. Each square is 1 square centimeter. Calculate the area of the cake box in square centimeters.

_____

**11.** The cake box is _____ cm wide and _____ cm high. Multiply these two, and place your answer here _____. Is it the same area as you calculated in question 10?

12. The cracker box above has a grid laid over it. Each square is 1 square inch. Calculate the area in square inches of the cracker box.

_____

13. The cracker box is _____ inches wide and _____ inches high. Multiply these two, and place your answer here _____. Is it the same area as you calculated in question 12?

**14.** The cookie box above has a grid laid over it. Each square is 1 square centimeter. Calculate the area in square centimeters of the cookie box.

_____

**15.** The cookie box is _____ cm wide and _____ cm high. Multiply these two, and place your answer here _____. Is it the same area as you calculated in question 14?

**16.** Lynn is having a picnic in the park. She spreads out her blanket as shown. What is the area it covers?

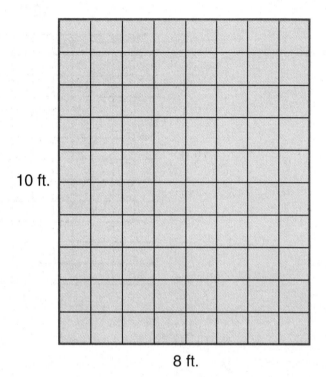

10 ft.

8 ft.

**A.** 18 ft²

**B.** 30 ft²

**C.** 60 ft²

**D.** 80 ft²

**17.** A wrestling mat is spread out in the gym for practice. What is the area of it?

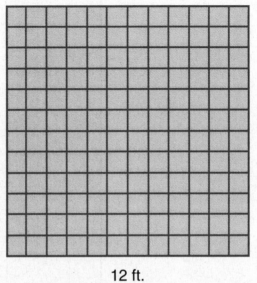

12 ft.

12 ft.                                    12 ft.

12 ft.

**A.** 120 ft²

**B.** 144 ft²

**C.** 164 ft²

**D.** 174 ft²

**18.** Hattie likes to display her jewelry collection on a felt cloth. What is its area?

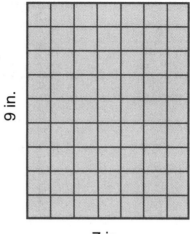

9 in.

7 in.

**A.** 63 in.$^2$

**B.** 60 in.$^2$

**C.** 74 in.$^2$

**D.** 70 in.$^2$

**19.** A coffee mug has a radius of 3 inches as shown. What is its (approximate) area?

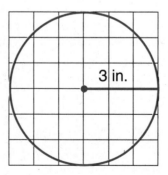

3 in.

**A.** 20 in.$^2$

**B.** 24 in.$^2$

**C.** 28 in.$^2$

**D.** 30 in.$^2$

**20.** A small window in the dining room is pictured below. What is its (approximate) area?

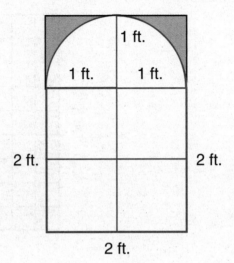

**A.** 5.5 ft²

**B.** 6 ft²

**C.** 4.5 ft²

**D.** 6.5 ft²

# AREA OF ODD SHAPES ON A SQUARE GRID

Some shapes are a combination of two or more rectangles or squares. When this happens, the best way to calculate the area is to break the shape into separate rectangles or squares. Then, the sum of the individual shapes can be added to obtain the total area of the shape.

For example, here is the floor area of a doll house:

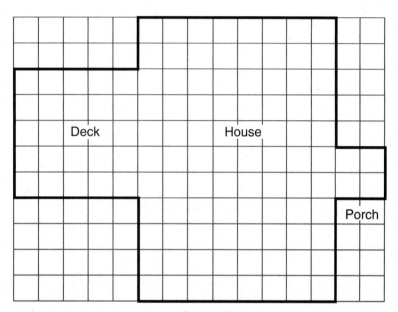

Scale: Each square is a square inch

In this example, the area is obtained by calculating the area of the three separate shapes (two squares and a rectangle) and adding up the three areas for the total area.

The deck is a square that is 5 inches on a side: $5 \times 5 = 25$ in.$^2$

The porch is a square that is 2 inches on a side: $2 \times 2 = 4$ in.$^2$

The house is a rectangle 11 inches long and 8 inches wide: $11 \times 8 = 88$ in.$^2$

The total area of the doll house is: $25 + 4 + 88 = 117$ in.$^2$

## AREA OF ODD SHAPES ON A SQUARE GRID EXERCISES

Answers appear on pages 215 and 216.

1. Courtney has a house that has a patio out back, and a side entrance with a porch on it, as diagrammed below (Note: squares are yards):

Scale: Each square is a square yard

**A.** What is the area of the house? _____ yd$^2$

**B.** What is the area of the patio? _____ yd$^2$

**C.** What is the area of the porch? _____ yd$^2$

**D.** What is the area of the whole house? _____ yd$^2$

**2.** Bob was tiling the floor of two rooms in a house. These rooms were both rectangles as the following figure illustrates (Note: squares are meters):

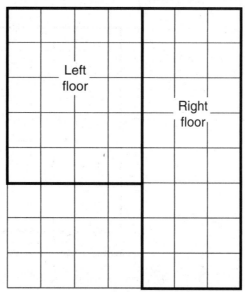

Scale: Each square is a square meter

**A.** What is the area of the left floor? _____ m²

**B.** What is the area of the right floor? _____ m²

**C.** What is the area of the two floors combined? _____ m²

3. Andrew was mowing a lawn and noticed that the front yard he was mowing was a square, and the side yard he was mowing was a rectangle, as illustrated in the following diagram (Note: squares are yards):

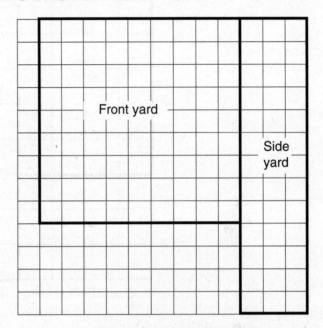

Scale: Each square is a square yard

A. What is the area of the front yard? _____ yd$^2$

B. What is the area of the side yard? _____ yd$^2$

C. What is the area of the two yards combined? _____ yd$^2$

**4.** A Victorian house often has wraparound porches. Here is an example (Note: squares are yards):

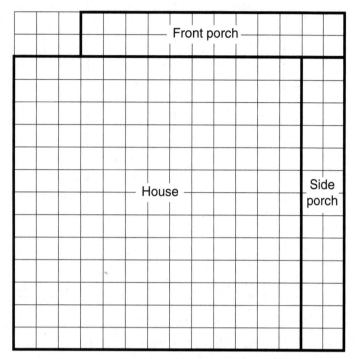

Scale: Each square is a square yard

**A.** What is the area of the house? _____ yd²

**B.** What is the area of the front porch? _____ yd²

**C.** What is the area of the side porch? _____ yd²

**D.** What is the area of the house and porches combined? _____ yd²

**5.** Angela was playing with blocks, and made a shape that combined three rectangles, as shown in the following diagram (Note: squares are centimeters):

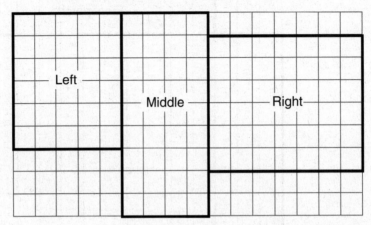

Scale: Each square is a square centimeter

**A.** What is the area of the left rectangle?
_____ cm²

**B.** What is the area of the middle rectangle?
_____ cm²

**C.** What is the area of the right rectangle?
_____ cm²

**D.** What is the area of the rectangles combined?
_____ cm²

# PROBLEM SOLVING AND OTHER MATHEMATICAL PROCESSES

## PROBLEM SOLVING

You solve problems every day. In fact, you solve problems every hour of every day. When you get up in the morning, you face a problem: "What will I wear?" The answer to that question requires the answer to at least two other questions:

1. What is in my closet that is clean?
2. What is the weather like today?

This is only one example of many problems you might face in the morning. Depending on your household, you might ask yourself, "What will I eat for breakfast?" or "Will I wash up first, then eat breakfast, or eat breakfast first, then wash up?" Not all people have these problems. Some people will insist on laying their clothes out the night before or following an unchanging routine every morning. But sometimes a change in routine is unavoidable (if someone is in the bathroom before you, for instance, or you have nothing clean the night before to lay out).

If you go on a school trip, you might consider what to bring (bottled water if you are outside on a hot day or a camera if there are opportunities to take pictures).

At lunch, you might ask yourself whether you want to eat indoors or out.

In the evening, you might consider whether you should do your homework before or after dinner.

A problem is a situation you face where you do not have a clear set of steps to solve it. It can also be a situation that you have never seen before. Facing a situation like that can be scary for many, but there are ways to approach it.

## POINTS TO KEEP IN MIND

You are not born with better or worse problem-solving ability than anyone else. Problem solving is a skill, not a talent. All skills are improved by practice, including problem solving. The best problem solvers get better with practice. Do not *ever* believe that you cannot solve problems, because you *can*. There are a few things you should remember as you attempt to solve a problem.

1. Believe that you can do it, because you *can*! If you believe you cannot do it, you will not even try. If you believe you can, you will try and try and try again until you succeed.
2. Do not think there is only one way to solve a particular problem. Very often there is more than one way to solve a particular problem or situation. We will look at a few strategies here.
3. Keep trying! You can solve it, if you keep at it. If you give up, you certainly will not solve it.
4. Look for a pattern in all solutions. Remember that mathematics is simply looking for patterns, making sure the pattern exists, and attempting to explain the pattern.

## METHOD FOR PROBLEM SOLVING

One of the most widely used methods for solving a problem has four steps:

1. **Make sense of the problem.** Understand all the aspects of the problem. Try to find out all the important information about the problem and to understand what the problem is asking. Think about what information (in the problem) is needed to solve the problem. What information (in the problem) is not needed to solve the problem? This might be the names of the people.

2. **Make a plan.** Once you understand all the aspects of the problem, you need to select the strategy you will use to solve the problem. Could you draw a picture? Could you make a chart or table? Could you think of a simpler problem? Could you guess the answer and then check it? Could you work backward from the end to the beginning? We'll look at each of these later on.

3. **Carry out the plan.** Now draw the picture or make up the table or write down the simpler problem or do the guessing and checking, to find the answer to the problem.

4. **Check your solution.** Now that you have an answer, stop! Look at the answer and think about it. Does it make sense? Does it answer the question completely? This last step is very important, but many people don't do it and give the wrong answer just because they didn't think about it. As an example of this, suppose you had to find out the age of a person. You thought of a plan, carried out the plan, and figured that the person is 900 years old! That of course makes no sense. People new to problem solving often will give an answer that makes no sense just because they don't stop to think about the answer.

---

***Note*** The other four process skills (Communication, Connections, Reasoning, and Representation) are addressed as the various questions are answered. For example, Communication and Representation are addressed when an answer is given, Reasoning goes on in the answering of the question, and Connections are addressed when we look at a problem involving a real-world problem, which all of these are.

Here are six problem-solving techniques (lettered A through F):

## A. Draw a Picture

If you are trying to tell someone how to get from your home to your school, you might draw a picture of the roads you need to take. Then you can remember what roads to look for and how many turns to make to get there. Very often, a picture or a diagram of some kind will show the pattern clearly.

### Example:

Davy wanted to have an 8-foot long pine board cut into eight pieces. The lumber yard charges $6.00 to cut a board into four pieces (and each cut costs the same). How much will the lumber yard charge Davy?

We will use the four-step process to solve this problem.

1. **Make sense of the problem.** We want to know how much Davy will be charged to have a board cut into eight pieces. The useless information is the fact that the wood is pine. We know how much it will cost for the lumber yard to cut it into four pieces. How can we solve this?

2. **Make a plan.** Drawing a picture would make the number of cuts clearer. That seems to be a good way to proceed. We will draw a picture of the board and show the number of cuts.

3. **Carry out the plan.** The board cut into four pieces looks like this:

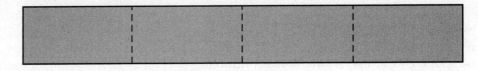

As we see, the board requires three cuts in it to be cut into four pieces. Since the lumber yard charges $6.00 to make three cuts, it is charging $$\$\frac{6}{3} = \$2 \text{ per cut.}$$

Now, we must extend this to seven cuts.

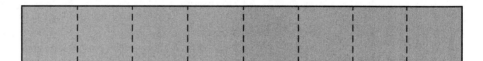

To cut the board into eight pieces, we use seven cuts, and since each cut is $2, the lumber yard will charge $2 × 7 = $14.

4. **Check your solution.** Now that we have the answer, let's check to see if it satisfies all the conditions of the problem. Look at the last figure. Is the board cut into eight pieces? Seven cuts gave us eight pieces. Then seven cuts times $2.00 per cut is $14.00. So it does satisfy the conditions of the problem.

## B. Make a Chart or a Table

You have a pocket full of change, and you want to know how much money you have. In this case, a table, where you write down how many quarters, dimes, nickels, and pennies you have, will help you find out how much money you have. Sometimes a chart, table, or diagram of some kind will reveal the pattern to the solution.

### Example:

Grace had pennies, nickels, dimes, and quarters in the pocket of her jumpsuit. If she reached in and pulled out two coins, how many different combinations could she have?

The four-step process:

1. **Make sense of the problem.** We want to find out how many different combinations of two coins Grace can make with four different kinds of coins. The coins are pennies, nickels, dimes, and quarters. The useless information is that she is wearing a jumpsuit. How can we solve this?

2. **Make a plan.** If we made a table, we could put down all the combinations. That way, we would avoid duplicate combinations, and make sure all combinations are given. We do not need to know how much each combination is worth in cents. It does not ask for that. A table seems to be the best way to proceed.

3. **Carry out the plan.** The table uses these abbreviations P = Penny, N = Nickel, D = Dime, and Q = Quarter and looks like this:

| Coins used | Combination number |
|:----------:|:------------------:|
| P P | 1 |
| P N | 2 |
| P D | 3 |
| P Q | 4 |
| N N | 5 |
| N D | 6 |
| N Q | 7 |
| D D | 8 |
| D Q | 9 |
| Q Q | 10 |

So, it would appear that there are 10 combinations of two coins that Grace can make.

4. **Check your solution.** Do 10 combinations seem reasonable? They do, and more than that, it appears that all the combinations are represented. Note that the table (or chart, if you like) makes it easy to check all the combinations, and see that there are no duplicates. Note also that order does not matter. That is, N Q is the same combination as Q N because the question simply asked for how many combinations, not the order of the coins.

## C. Think of a Simpler Problem

You go to a party with six other people (besides yourself). How many handshakes does it take for all the people to shake everyone else's hand? Oftentimes a problem has a pattern that can be seen in smaller versions of the problem and can then be built up to the larger problem.

### Example:

Gill built a stair with wooden building blocks. They had letters of the alphabet on them. How many blocks does he need to build a stair that has eight steps to it.
  The four-step process:

1. **Make sense of the problem.** We need to find out how many blocks it will take to build a stair with eight steps to it. The useless information seems to be that the blocks have alphabet letters on them. How can we solve this?
2. **Make a plan.** A stair with eight steps seems pretty big. If we look at a smaller stair, we can find the number of blocks easily, and thereby see the pattern that is formed.

**3. Carry out the plan.** If Gill makes a stair with one step, it will take just one block.

If he makes a stair of two steps, it will take three blocks.

If he makes a stair of three steps, it will take six blocks.

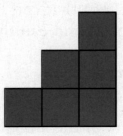

If he makes a stair of four steps, it will take 10 blocks.

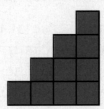

If he makes a stair of five steps, it will take 15 blocks.

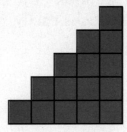

If he makes a stair of six steps, it will take 21 blocks.

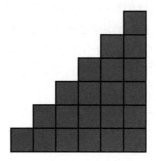

If he makes a stair of seven steps, it will take 28 blocks.

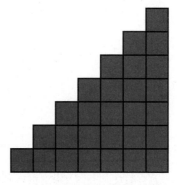

Finally, if he makes a stair of eight steps, it will take 36 blocks.

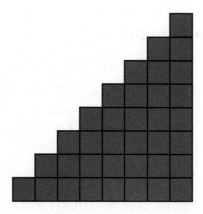

4. **Check your solution.** Clearly, the figures show that a stair of eight steps will take 36 blocks. Starting with a simpler problem helps us to build up to the solution. Another way, though, might be to use a table, as with the last problem:

| Steps | Blocks |
|-------|--------|
| 1     | 1      |
| 2     | 3      |
| 3     | 6      |
| 4     | 10     |
| 5     | 15     |
| 6     | 21     |
| 7     | 28     |
| 8     | 36     |

Again, this clearly shows how, starting with a simpler problem, the solution to a larger problem can be easily solved.

## D. Try, Guess, and Check

You want to put your collection of 28 model cars into four garages. Can you do it? Sometimes you can get an answer just by guessing it. It usually needs to be checked, though, to see if it fulfills the conditions of the problem.

### Example:

Elena has seven friends from ballet class over to her house. She wants to give cookies to each of the seven friends. She has 30 cookies, and she wants to keep two for herself. Can she offer each friend the same number of cookies with none left over?

The four-step process:

1. **Make sense of the problem.** We want to find out if there is a number of cookies Elena can give her friends with two cookies left over for her. The useless information is that these friends are in ballet class together. What can we do to solve this?

2. **Make a plan.** One way of solving this is to guess the answer and then check the solution against the conditions of the problem. That seems like a good plan.

3. **Carry out the plan.** We will first remember that Elena wants two cookies for herself. That means that she should take away two cookies from the total: $30 - 2 = 28$. She needs to distribute 28 cookies to all her friends. Let's try giving the same number of cookies to her friends that she took: two.

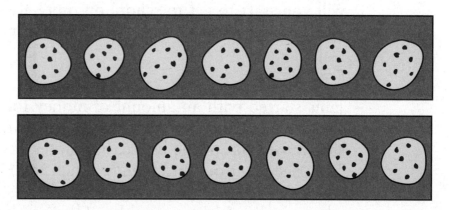

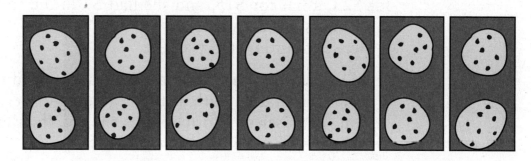

As you can see, if she give two cookies to each person, she will have 14 cookies left. But that is just the number we used in our first guess. So let's give another two cookies to each friend, like this:

Now, we have it. We have used up all the cookies and each friend has the same number of cookies: four.

4. **Check your solution.** We see that we have the answer. Each friend can get four cookies with no cookies left over. We guessed the answer, then checked it. It does seem reasonable.

## E. Work Backward

You want to plan to get to school on time. School starts at 8:00 A.M. You take 10 minutes to get to school. Breakfast takes 30 minutes to make and eat, and it takes 25 minutes for you to get dressed for school. What time will you start to get to school on time? Remembering each step will help you get back to what you started with.

**Example:**

Janis started with an amount of money for shopping, but she doesn't remember what she started with. Just before she walked out of the house, her father gave her $20. Then she went out and bought lunch for $7. She bought a belt for $18. While paying for the belt, she met a friend who gave her $15 he owed her. She then bought a blouse for $23, a hat for $15, and she had $7 in the end. How much did she start with?

Here is the four-step process:

1. **Make sense of the problem.** We want to find out how much she started with at the beginning of her shopping trip. The useless information is the specific things she bought. What will we do to solve this?

2. **Make a plan.** If we start with the final amount, and work backward, doing the opposite of what Janis did, we will end up with what she started with. So, for instance, if she spends money, we will add that amount to the total. If she gets money, we will subtract it from her total. That way, we will back up to the original amount. So that seems like a good plan.

3. **Carry out the plan.** We will set up a running total, adding or subtracting as needed. Janis ended with $7, so that is what we will start off with.

She had $7 just before she bought a hat for $15, so

$$\begin{array}{r} \$7 \\ +\$15 \\ \hline \$22 \end{array}$$

She had $22 before she bought the hat. She bought a blouse for $23 before that, so

$$\begin{array}{r} \$22 \\ +\$23 \\ \hline \$45 \end{array}$$

She had $45 before the blouse. She received $15 from a friend before that, so

$$\begin{array}{r} \$45 \\ -\$15 \\ \hline \$30 \end{array}$$

She had $30 before she met her friend. She bought a belt for $18 before that, so

$$\begin{array}{r} \$30 \\ +\$18 \\ \hline \$48 \end{array}$$

She had $48 before she bought the belt. She bought lunch before the belt for $7, so

$$\begin{array}{r} \$48 \\ +\$\ 7 \\ \hline \$55 \end{array}$$

She had $55 before she got lunch. She was given $20 before that, so

$$\begin{array}{r} \$55 \\ -\$20 \\ \hline \$35 \end{array}$$

She had $35 before her father gave her $20, so that's what she started with.

4. **Check your solution.** To check the solution, use the starting amount we found, and do the same operations on it that the problem gives:

$$\$35 + 20 = \$55$$
$$\$55 - \$7 = \$48$$
$$\$48 - \$18 = \$30$$
$$\$30 + \$15 = \$45$$
$$\$45 - \$23 = \$22$$
$$\$22 - \$15 = \$\ 7$$

$7 is what Janis ended up with, so this does check out.

## F. Simulate It

That is, use pictures or things to act it out. You have 21 students in your class, and you are going on a field trip. Your school has vans that carry seven students each. How can you divide up the class? Using beans and dividing them up can help you see the solution to this one.

**Example:**

Anastasia has a collection of 48 small stuffed animals. She wants to put them on her bookcase. The bookcase has lots of shelves, and each shelf holds six stuffed animals. How many shelves does she need to arrange them?

1. **Make sense of the problem.** We need to figure out how many shelves we will need to arrange the animals. We will be arranging them in groups of six. How can we solve this?

2. **Make a plan.** If we picture them in groups of six, we can find out how many shelves we need. So drawing a picture or using beans or some other manipulative seems like a good way to solve this.

3. **Carry out the plan.** We draw the picture that arranges the stuffed animals in groups of six until all the stuffed animals are used up:

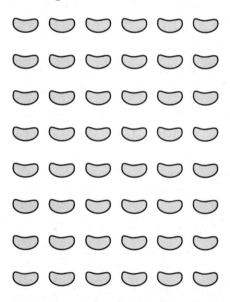

As you can see, we can arrange the stuffed animals on eight shelves of six each. This arrangement will store all the animals, with no partially full shelves. Using mathematical language, this is

$$8 \times 6 = 48$$

4. **Check your solution.** Did we store all 48 of the stuffed animals? Yes, and there were no shelves that were partially filled with stuffed animals. We used eight shelves to store the stuffed animals six to a shelf. Another way to solve this would be to start with 48 beans, and take six away at a time, keeping track of how many times we take six away.

## PROBLEM-SOLVING EXERCISES

Answers appear on pages 216 to 220.

Workspace has been provided for you to use when solving the problems below.

1. Greg is having several friends over for a barbeque in the afternoon. He will be cooking hamburgers, hot dogs, ribs, and chicken drumsticks on the grill. Hamburgers take 20 minutes to cook. Ribs take 35 minutes to cook. Hot dogs take 10 minutes to cook. Chicken drumsticks take 30 minutes to cook. He wants to take all the items off the grill at 3:00 P.M. If he has a grill that will fit all the different meats he is cooking, when should he put each meat on the grill to ensure that they will all come off at 3:00 P.M.?

**2.** Some members of the Gaskin family (including some aunts and uncles) went to River City Zoo recently. The children invited some of their friends, so that there were more children than adults. They paid $104 to get the whole family in. If the admission price for children is $7, and the admission price for adults is $12, how many children and how many adults went on the Gaskin's zoo trip?

**3.** Phil and Ben took a kayak trip in Long Lake. They paddled away from Hale Dock at 8:00 A.M., and went east, paddling 5 mph. They went this way until noon, at which time they pulled onto shore for lunch. Then they got back into their kayaks at 1:00 P.M. and paddled west at a speed of 4 mph, paddling until 4:00 P.M. How close were they to the dock by then?

4. For Earth Day, Andrew and his brother Henry are picking up plastic bottles and cans along the Passaic River. Andrew picks up six bottles/cans for every five bottles/cans that Henry picks up. They work 2 hours and together they collect 132 bottles/cans. How many bottles/cans did each boy pick up?

5. A triangular array of dots looks like this:

How many dots would there be in a triangular array having 10 dots on each side?

## PROBLEM SOLVING AND OTHER MATH SKILLS

Answers appear on pages 221 to 230.
Use the workspace that has been provided.

**1.** In a sequence of steps, we know all of them except the
first one. What is the first number?

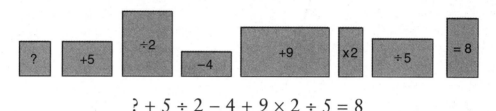

$$? + 5 \div 2 - 4 + 9 \times 2 \div 5 = 8$$

**2.** Zack wants to put a fence around his mom's vegetable garden. The garden is in one corner of the yard and is shaped like a trapezoid:

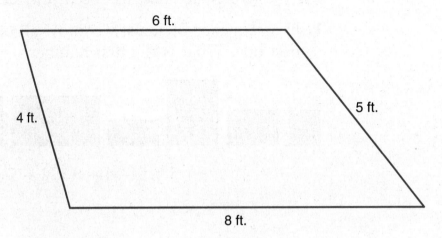

How many feet of fence will Zack need to get?

**3.** The lockers on the first floor in Third River High School are numbered 100–200. How many lockers have a 6 in them?

**4.** Consider this number sequence: 23, 28, 26, 31, 29, 34,...

What is the next number in this sequence?

**5.** Peter wants to plant a tree in his front yard. The yard is 50 feet wide. The dirt ball around the roots of the tree is 4 feet. How far from each side of the yard must he plant the tree to center it in the front yard?

6. Mickey has a quarter, a nickel, and two pennies in his pocket. How many different sums of money can he make?

7. The battleship *New Jersey* gives a break to school groups. For every eight tickets sold, the group gets one free ticket. How many free tickets will a school get if there are 50 in the school group?

**8.** Consider the following graph:

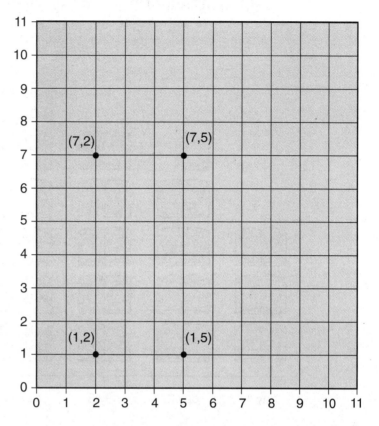

**A.** If these four points were to be connected, what two-dimensional geometric shape would be formed?

**B.** What is the perimeter of the shape?

**C.** If you were to slide this shape four units up, where would the points be plotted on the graph?

**9.** The following pictograph shows the amount of books sold at Bob and Ray's Book Shop over the last 6 days:

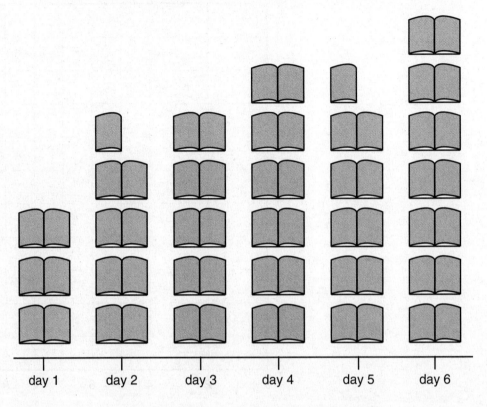

Each book represents 10 books

**A.** How many books did Bob and Ray's Book Shop sell on the fourth day?

**B.** On what day did the sales go down?

**C.** How many books did Bob and Ray's Book Shop sell for the entire 6 days?

**10.** Martina is getting outfits together for her Caribbean vacation. She has four blouses (one black, one gray, one blue, and one white), and three pairs of pants (one black, one dark gray, and one white). How many different outfit combinations can she make for herself? Make up a tree or other diagram to show your solution.

If she had two pairs of shoes (one sneakers and one penny loafers), how many blouses, pants, and shoes combinations can she make? Use a tree or other diagram to show your solution.

# ANSWERS TO PRACTICE PROBLEMS

## CHAPTER 2

### ANSWERS TO MULTIPLICATION FACT EXERCISES

1. **A.** (Hint: if you count by 5s seven times you will see the answer is 35).

2. **B.** (Hint: if you count by 2s nine times you will see the answer is 18).

3. **A.** (If you do the nines trick using your fingers you will see the answer is 63).

4. **B.** ($7 \times 7$ is 49. You could memorize this fact, or create an array to solve it).

5. **C.** (Any number times 0 is 0).

### ANSWERS TO DIVISION FACT EXERCISES

1. **B.** (In the problem $42 \div 6$ there are a total of seven 6s in the number 42).

2. **A.** (In the problem $25 \div 5$ there are a total of five 5s in the number 25).

3. **D.** (In the problem $63 \div 7$ there are a total of nine 7s in the number 63).

4. **A.** (In the problem $30 \div 5$ there are a total of six 5s in the number 30).

**5. C.** (In the problem 36 ÷ 9 there are a total of four 9s in the number 36).

## ANSWERS TO MISSING VARIABLE AND DIVISION/MULTIPLICATION FACT FAMILY EXERCISES

**1. B.** (In this fact family ? = 7. We are solving 28 ÷ 4 = 7, or alternately 7 × 4 = 28).

**2. A.** (In this fact family $r$ = 7. We are solving 42 ÷ 6 = 7, or alternately 7 × 6 = 42).

**3. D.** (In this fact family $b$ = 36. We are solving 36 ÷ 6 = 6, or alternately 6 × 6 = 36).

**4. C.** (In this fact family $x$ = 16. We are solving 16 ÷ 2 = 8, or alternately 2 × 8 = 16).

**5. A.** (In this fact family $s$ = 9. We are solving 72 ÷ 8 = 9, or alternately 8 × 9 = 72).

**6.** The four facts for this fact family are:

$$7 \times 8 = 56$$

$$8 \times 7 = 56$$

$$56 \div 7 = 8$$

$$56 \div 8 = 7$$

**7.** Before listing the four facts we must first solve for *r*. Since the product of 63 is given, and one factor is 9, we know that the other factor must be 9 since 9 × 7 = 63. Now we can proceed to list the four facts for this fact family:

$$9 \times 7 = 63$$

$$7 \times 9 = 63$$

$$63 \div 7 = 9$$

$$63 \div 9 = 7$$

**8.** Before listing the four facts we must first solve for *t*. Since the two factors are given (4 and 8) we can simply multiply to find the missing value (we know this must be the product because it is notated with the dot). 4 × 8 = 32, which is our missing value. Now we can proceed to list the four facts for this fact family:

$$4 \times 8 = 32$$

$$8 \times 4 = 32$$

$$32 \div 4 = 8$$

$$32 \div 8 = 4$$

## ANSWERS TO SINGLE-STEP WORD PROBLEMS

1.

| Piles | Books per pile | Books in all |
|-------|----------------|--------------|
| 4     | ?              | 24           |

Answer: _____6_____ books per pile
(unit)

If there are 24 books altogether, and they are split between four piles, then this is a division problem: 24 ÷ 4 = 6. You can utilize some of the methods taught to determine the answer of 6 if you do not have the facts memorized. Examples would be building an array, or sorting into groups.

2.

| Rows | Cars per row | Cars in all |
|------|--------------|-------------|
| 6    | 8            | ?           |

Answer: _____48_____ cars in all
(unit)

In this problem we have six rows with eight in each. This is a basic multiplication problem as we are counting by 8s six times (alternately, we could also count by 6s eight times). If you do not know the answer for 6 × 8 you could build an array 6 × 8 and count them up to arrive at an answer of 48.

**3.**

| Days | Classes per day | Classes in all |
|------|-----------------|----------------|
| 4 | 3 | ? |

Answer: _____ 12 _____ classes in all
           (unit)

In this problem, Jamie is teaching three classes per day, and she is doing this four times each week. This is a basic multiplication problem as we are counting by 3s four times (alternately, we could also count by 4s three times). If you do not know the answer for $4 \times 3$ you could build an array $4 \times 3$ and count them up to arrive at an answer of 12.

**4.**

| People | Tacos per person | Tacos in all |
|--------|------------------|--------------|
| 8 | 2 | ? |

Answer: _____ 16 _____ tacos in all
           (unit)

This problem is a little tricky because you must remember to include Jennifer. She is also having tacos. With her included there are eight people and each is having two tacos. This again is a multiplication problem. You can count by 8s two times, or count by 2s eight times. In either event, you are solving $8 \times 2 = 16$.

**5.**

| Vans | Students per van | Students in all |
|------|------------------|-----------------|
| ? | 6 | 24 |

Answer: _____4_____ vans
(unit)

This is a division problem. If we know there are 24 students altogether, and they must be split into quantities (or groups) of six to fit in vans, then we are solving 24 ÷ 6. Our lesson on fact families taught us that this is really asking what number times 6 equals 24. Thinking through your six facts you should come up with an answer of four. If you did not memorize your facts yet you could take the 24 and split them into equal groups of six. Then you would see that there are four groups (vans) in total.

## ANSWERS TO TWO-STEP WORD PROBLEMS

1. The answer is $4.00. If we count up all the fruit that Ron purchased we get 40 (6 + 3 + 11 + 20 = 40). Now that we have the total number of fruits, we need to multiply this by $0.10 to determine the total cost. 40 × $0.10 is $4.00.

2. The answer is 14 bottles. If Mark drinks a bottle per hour at the park, and he is there for 2 hours, we need to multiply 1 × 2 to determine how many bottles are needed each day at the park. Since two bottles are needed each day, we then must multiply this by 7 (the number of days they go to the park). 2 × 7 = 14, so Marilyn needs to purchase 14 bottles of water to meet Mark's needs each week at the park.

**3.** The answer is $120.00. First we need to determine the total number of hours Carly babysat. To do this, we add up all of her hours: $2 + 3 + 2 + 4 + 5 + 3 + 1 = 20$. Now that we know she babysat for a total of 20 hours last month, we can multiply this by how much money she is paid each hour ($6.00). $20 \times \$6.00 = \$120.00$. So last month Carly made $120.00 babysitting for her neighbor.

**4.** The answer is 36 hours. We need to first figure out how many hours Tom mows each week. The problem tells us he mows three lawns per week. To get the total, we just add up the number of hours it takes to mow each lawn each week. $3 + 2 + 4 = 9$. So each week Tom spends 9 hours mowing lawns. However, the problem is asking how many hours he works in a month (4 weeks), so we just have to multiply 9 (hours per week) $\times$ 4 (weeks) = 36 hours.

**5.** The answer is 22 items. When Yanni went to the store, he originally purchased a total of 24 items $(6 + 11 + 7 = 24)$. When he got home and realized he already had two of the items, he returned them to the store. So Yanni now has two less items than he originally did when he left the store. $24 - 2 = 22$. Yanni ended up with 22 new items.

## ANSWERS TO PATTERNS PRACTICE

**1.** The next four numbers in the array would be 17, 20, 23, and 26.

The pattern is that the next number in the array is increased by 3.

2. The next three numbers in the array would be 230, 221, 211, and 200.

   The pattern is that the next number in the array is decreased by one more each time. (The difference between the first and the second number is 5. The difference between the second and third numbers is 6; the difference between the third and fourth numbers is 7, and so forth.)

3. In this example, we are given the rule of × 6. We can simply apply this rule to our given "in" numbers to obtain the "out" numbers. 6 × 6 = 36 and 4 × 6 = 24. For the missing "in" numbers we need to think of it as _____ × 6 = 30 or we can think of it in terms of division: 30 ÷ 6 = _____. In either case, the answer is 5. Similarly we can apply it to the 6. Think of it as 6 ÷ 6 = _____, or _____ × 6 = 6. In both cases, the answer is 1.

| RULE |
| --- |
| × 6 |

| IN | OUT |
| --- | --- |
| 6 | 36 |
| 4 | 24 |
| 5 | 30 |
| 1 | 6 |
| | |

4. In this example, we are given the rule of ×4. We can simply apply this rule to our given "in" numbers to obtain the "out" numbers. $9 \times 4 = 36$ and $0 \times 4 = 0$. For the missing "in" numbers we need to think of it as _____ $\times 4 = 28$ or we can think of it in terms of division: $28 \div 4 =$ _____. In either case the answer is 7. Similarly we can apply it to the 4. Think of it as $4 \div 4 =$ _____, or _____ $\times 4 = 4$. In both cases the answer is 1.

| RULE |
| --- |
| × 4 |

| IN | OUT |
| --- | --- |
| 7 | 28 |
| 1 | 4 |
| 9 | 36 |
| 0 | 0 |
|  |  |

5. In this example, we are given all of the "in" and "out" values and we are asked to determine the rule. To determine the rule we need to pick a row and figure out what is happening. Let us look at the 3 and the 6. At first glance you might think + 3. However, *always* check another row to be certain. Let us look at the 22 and the 44. We can quickly see that the rule is not + 3 because it does not work on this row. Instead, the rule is *double* (or think of it as × 2).

| RULE |
| --- |
| Double |

| IN | OUT |
| --- | --- |
| 3 | 6 |
| 9 | 18 |
| 22 | 44 |
| 14 | 28 |
|  |  |

**6.** In this example, we are given the rule of ÷ 2. We can simply apply this rule to our given "in" numbers to obtain the "out" numbers. 6 ÷ 2 = 3 and 20 ÷ 2 = 10. For the missing "in" numbers we need to think of it as _____ ÷ 2 = 8 or we can think of it in terms of multiplication: 8 × 2 = _____. In either case the answer is 16. Similarly we can apply it to the 5. Think of it as _____ ÷ 2 = 5, or 5 × 2 = _____. In both cases the answer is 10.

| RULE |
|------|
| ÷ 2 |

| IN | OUT |
|----|-----|
| 6 | 3 |
| 16 | 8 |
| 10 | 5 |
| 20 | 10 |
| | |

7. In this example, we are given the rule of +4. We can simply apply this rule to our given "in" number to obtain the "out" number. 28 + 4 = 32. For the missing "in" numbers we need to think of it as _____ + 4 = 7 or we can think of it in reverse using subtraction: 7 – 4 = _____. Either way we get an answer of 3. If we apply this same thinking to the 47 we get 43: _____ + 4 = 47, or 47 – 4 = _____. Finally, for the 61 we get an answer of 57: _____ + 4 = 61, or 61 – 4 = _____.

| RULE |
|------|
| + 4 |

| IN | OUT |
|----|-----|
| 3 | 7 |
| 28 | 32 |
| 43 | 47 |
| 57 | 61 |
| | |

## CHAPTER 3

### ANSWERS TO PLACE VALUE EXERCISES

1. **B.** (The 0 is the third digit from the right, which is the hundreds position).

2. **C.** (The 8 is the first digit from the right, which is the ones position).

3. **D.** (The 8 is the fourth digit from the right, which is the thousands position).

4. **A.** (The 1 is the fourth digit from the right, which is the thousands position).

5. **D.** (The 4 is the third digit from the right, which is the hundreds position).

### ANSWERS TO ESTIMATION EXERCISES

1. **C.** (587 is closer to 600 than it is to 500, if you were rounding to the nearest hundred).

2. **D.** (3,971 is closer to 4,000 than it is to 3,000, if you were rounding to the nearest thousand).

3. **B.** (32,800 is the only number that is rounded to the hundreds position. In this number, the 8 is in the hundreds position and every other digit to the right is 0).

4. **A.** (76,000 is the only number that is rounded to the hundreds position. In this number, the 6 is in the thousands position and every other digit to the right is 0).

5. **D.** (3,971 is closer to 4,000 than it is to 3,000 if you were rounding to the nearest thousand).

6. **C.** (Rounding to the tens: 86 = 90, 74 = 70, 84 = 80, and 70 = 70. Because the question is asking which one would be 80 when rounded to the tens, we can see the answer is C).

7. **A.** (Rounding to the hundreds: 944 = 900, 849 = 800, 951 = 1,000, and 989 = 1,000. Because the question is asking which one would be 900 when rounded to the hundreds, we can see the answer is A).

8. **D.** (1,600 is already rounded to the hundreds; 1,632 would be 1,600; 7,789 would round to 7,800; and 1,656 would round to 1,700. Because the problem asks which number rounded to the hundreds has a 7 in the hundreds position, we can see the answer is D).

## ANSWERS TO ADDITION EXERCISES

1. Answer = C

$$
\begin{array}{r}
{}^{1}\phantom{0} \\
5\ 8 \\
+\ \phantom{0}9 \\
\hline
6\ 7
\end{array}
$$

In this problem, 8 + 9 = 17. Bring down the 7, carry the 1. 1 + 5 = 6. So the answer is 67.

2. Answer = A

$$
\begin{array}{r}
{}^{1}\phantom{00} \\
3\ 4\ 1 \\
+\ 5\ 0\ 9 \\
\hline
8\ 5\ 0
\end{array}
$$

In this problem, 9 + 1 = 10. Bring down the 0, carry the 1. 1 + 4 + 0 = 5. 3 + 5 = 8. So the answer is 850.

3. Answer = B

$$
\begin{array}{r}
4\ 1\ 3 \\
+\ 3\ 0\ 1 \\
\hline
7\ 1\ 4
\end{array}
$$

**4.** Answer = A

$$
\begin{array}{r}
\overset{1}{\phantom{0}}\overset{1}{\phantom{0}}\phantom{0} \\
5\ \ 4\ \ 3 \\
3\ \ 9\ \ 9 \\
+\ \ \ \ \ 4\ \ 2 \\
\hline
9\ \ 8\ \ 4
\end{array}
$$

**5.** Answer = D

$$
\begin{array}{r}
\overset{1}{\phantom{0}}\phantom{00} \\
4\ \ 1 \\
3\ \ 2 \\
9\ \ 9 \\
+\ \ 1\ \ 4 \\
\hline
1\ \ 8\ \ 6
\end{array}
$$

## ANSWERS TO SUBTRACTION EXERCISES

**1.** Answer = B

$$
\begin{array}{r}
6\ \ 2\ \ 3 \\
-\ \ 2\ \ 1\ \ 1 \\
\hline
4\ \ 1\ \ 2
\end{array}
$$

This problem did not involve any borrowing because the number being subtracted each time was smaller than the original (top) number.

**2.** Answer = A

$$
\begin{array}{r}
\phantom{0}\overset{3}{\phantom{0}}\ \overset{13}{\phantom{0}} \\
8\ \ \cancel{4}\ \ \cancel{3} \\
-\ \ 1\ \ 0\ \ 9 \\
\hline
7\ \ 3\ \ 4
\end{array}
$$

In this problem we had to borrow from the 4 because we could not do 3 − 9.

3. Answer = C

$$
\begin{array}{ccc}
 & \overset{4}{\cancel{5}} & \overset{12}{\cancel{2}} \\
7 & & \\
- \quad 2 & 2 & 9 \\
\hline
5 & 2 & 3 \\
\end{array}
$$

In this problem we had to borrow from the 5 because we could not do 2 – 9.

4. Answer = D

$$
\begin{array}{ccc}
 & \overset{7}{\cancel{8}} & \overset{10}{\cancel{0}} & 7 \\
- & 1 & 4 & 2 \\
\hline
 & 6 & 6 & 5 \\
\end{array}
$$

In this problem we had to borrow from the 8 because we could not do 0 – 4.

5. Answer = C

$$
\begin{array}{ccc}
 & & \overset{9}{} & \\
 & \overset{8}{\cancel{9}} & \overset{\cancel{10}}{\cancel{0}} & \overset{10}{\cancel{0}} \\
- & & 6 & 5 \\
\hline
 & 8 & 3 & 5 \\
\end{array}
$$

This is by far the trickiest problem of them all. In this problem, we have to borrow twice. If we start on the right (as always) we see that we cannot do 0 – 5. Looking to the left is another 0 so we cannot borrow from it. Therefore, we move again to the left and borrow from the 9. The 9 becomes an 8 and the middle 0 becomes a 10. Now that there is a 10 there we can cross it off and borrow from it to help the original 0. The 10 becomes a 9 and the original 0 (all the way to the right) becomes a 10. Now we can subtract. 10 – 5 = 5; 9 – 6 = 3; 8 – 0 = 8. The final answer is 835.

## ANSWERS FOR MULTIPLICATION BY MULTIPLES OF 10 EXERCISES

**1.** Answer = B

We can solve this using the fact extension method, which would be $7 \times 9$ (instead of $7 \times 90$), which is 63, then put the 0 back onto the end and we come up with 630. We can also solve using standard multiplication:

| STEP 1<br>Rewrite vertically | STEP 2<br>Multiply by ones position | STEP 3<br>Multiply by tens position |
|:---:|:---:|:---:|
| 9 0<br>× 7 | 9 0<br>↑<br>× 7<br>―――<br>0 | 9 0<br>↖<br>× 7<br>―――<br>6 3 0 |

**2.** Answer = B

We can solve this using the fact extension method which would be $8 \times 5$ (instead of $8 \times 50$), which is 40, then put the 0 back onto the end and we come up with 400. We can also solve using standard multiplication:

| STEP 1<br>Rewrite vertically | STEP 2<br>Multiply by ones position | STEP 3<br>Multiply by tens position |
|:---:|:---:|:---:|
| 8 0<br>× 5 | 8 0<br>↑<br>× 5<br>―――<br>0 | 8 0<br>↖<br>× 5<br>―――<br>4 0 0 |

**3.** Answer = C

We can solve this using the fact extension method which would be $9 \times 1$ (instead of $9 \times 10$) which is 9, then put the 0 back onto the end and we come up with 90. We can also solve using standard multiplication:

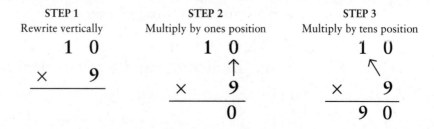

| STEP 1<br>Rewrite vertically | STEP 2<br>Multiply by ones position | STEP 3<br>Multiply by tens position |
|:---:|:---:|:---:|
| 1 0<br>× 9 | 1 0<br>↑<br>× 9<br>―――<br>0 | 1 0<br>↖<br>× 9<br>―――<br>9 0 |

**4. Answer = D**

We can solve this using the fact extension method, which would be $7 \times 7$ (instead of $7 \times 70$) which is 49, then put the 0 back onto the end and we come up with 490. We can also solve using standard multiplication:

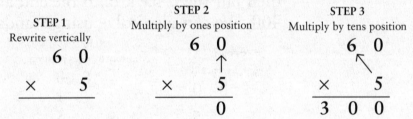

| STEP 1<br>Rewrite vertically | STEP 2<br>Multiply by ones position | STEP 3<br>Multiply by tens position |
|---|---|---|
| 7 0<br>× 7 | 7 0<br>↑<br>× 7<br>0 | 7 0<br>↖<br>× 7<br>4 9 0 |

**5. Answer = A**

We can solve this using the fact extension method, which would be $5 \times 6$ (instead of $5 \times 60$) which is 30, then put the 0 back onto the end and we come up with 300. We can also solve using standard multiplication:

| STEP 1<br>Rewrite vertically | STEP 2<br>Multiply by ones position | STEP 3<br>Multiply by tens position |
|---|---|---|
| 6 0<br>× 5 | 6 0<br>↑<br>× 5<br>0 | 6 0<br>↖<br>× 5<br>3 0 0 |

# CHAPTER 4

## ANSWERS TO FRACTION EXERCISES

1. C. (five out of eight pieces are shaded in the diagram. The denominator is 8 because that was the total number of pieces. The numerator is 5 because it represents five shaded pieces).

2. C. There are eight pieces in all. 3 are not shaded. So 3 is our numerator and 8 is our denominator.

3. **D.** (The denominator in this fraction is 6 because it represents the total number of segments. The numerator is 1 because it represents the value that one segment has on the entire diagram. Hence, one segment represents $\frac{1}{6}$, or one-out-of six possible).

4. **C.** (The denominator in this fraction is 3 because it represents the total number of segments. The numerator is 2 because it represents two segments, which is what the question asked for. Hence, the distance of two segments combined is equal to $\frac{2}{3}$ of the entire line. In other words, it is two-out-of-three possible).

5. **C.** (Because the given denominators are all 5, that tells us this diagram is split into fifths. In the second missing space, we notice that it is directly to the left of the $\frac{3}{5}$ that the diagram gives us. From this, we can just count backwards by one and come up with the fraction $\frac{2}{5}$. The denominator does not change because it is still referring to the total of five segments in the diagram. The numerator changed from 3 to 2 because the distance traveled from the start is now one less, so it needs to be one less.

## ANSWERS TO EQUIVALENT FRACTIONS EXERCISES

1. **B.** (Using cross-multiplication we calculate $4 \times 2 = 8$ and $3 \times 3 = 9$. Since 9 is larger than 8, this tells us that the fraction on that side is larger.

$$\overset{8}{}\frac{2}{3} \,\,\bcancel{\times}\,\, \frac{3}{4}\overset{9}{}$$

**2. A.** (Using cross-multiplication we calculate $3 \times 3 = 9$ and $4 \times 1 = 4$. Since 9 is larger than 4, this tells us that the fraction on that side is larger.

$$\overset{9}{\frac{3}{4}} \bowtie \overset{4}{\frac{1}{3}}$$

**3. C.** (Using cross-multiplication we calculate $6 \times 2 = 12$ and $2 \times 6 = 12$. Because 12 is equal to 12, this tells us that the two fractions are equivalent.

$$\overset{12}{\frac{2}{2}} \bowtie \overset{6}{\frac{3}{6}}$$

**4. C.** (Using cross-multiplication we calculate $6 \times 4 = 24$ and $8 \times 3 = 24$. Because 24 is equal to 24, this tells us that the two fractions are equivalent.

$$\overset{24}{\frac{4}{8}} \bowtie \overset{24}{\frac{3}{6}}$$

**5. C.** (There are five shaded pieces out of a total of six pieces. The shaded region represents $\frac{5}{6}$ of the entire circle.)

**6. A.** (There is one unshaded piece out of a total of four pieces. The unshaded region represents $\frac{1}{4}$ of the entire circle).

**7. D.** (There are two shaded pieces out of a total of three pieces. The shaded region represents $\frac{2}{3}$ of the entire circle).

8. **A.** (Since we are trying to find a fraction that is equivalent, it means that when we have two choices, we can either compare the numerators and figure out the rule (×3, ×2, etc.), and then see if it applies to the denominator as well, or we can do cross-multiplication. If you choose cross-multiplication this is a trial-and-error problem so you really just need to pick an answer, do the math, and see if they come out equal.

$$\overset{24}{\frac{3}{4}} \bowtie \overset{6\,24}{\frac{6}{8}}$$

Alternately, you can see that the rule here is ×2 if you compare the numerators. You start out with a 2 and end up with a 6. If the ×2 rule works on the denominators as well, then the fraction is equivalent. In this case, it does work because $4 \times 2 = 8$.

9. **D.** (Since we are trying to find a fraction that is equivalent, it means that when we have two choices, we can either compare the numerators and figure out the rule (×3, ×2, etc.), and then see if it applies to the denominator as well, or we can do cross-multiplication. If you choose cross-multiplication this is a trial-and-error problem so you really just need to pick an answer, do the math, and see if they come out equal.

$$\overset{?}{\frac{5}{8}} \bowtie \overset{15\,?}{\frac{15}{24}}$$

It's easy to see that cross-multiplication is going to be quite complex in this problem, so skip it and try comparing the numerators instead. If we start out as a 5 and turn into a 15 the rule is ×3. So if ×3 works on the denominators also then these are equivalent fractions. Because $8 \times 3 = 24$, these are equivalent fractions.

10. **A.** (There are three unshaded pieces out of a total of eight pieces. The unshaded region represents $\frac{3}{8}$ of the entire circle).

## CHAPTER 5

### ANSWERS TO CUSTOMARY (U.S. STANDARD) SYSTEM EXERCISES

1. **C.** The stamp is $1\frac{1}{2}$ inches wide, which is closest to $1\frac{3}{4}$ inches.

2. **A.** The MP3 player is 3 inches long.

3. **D.** The cities are miles apart.

4. **C.** The ship is 4 inches long, and 3 inches high.

5. **B.** A yard is 3 feet.

6. The coins line graph follows:

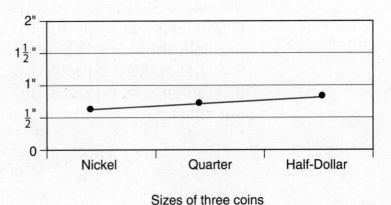

Sizes of three coins

**7.** The stamps line graph follows:

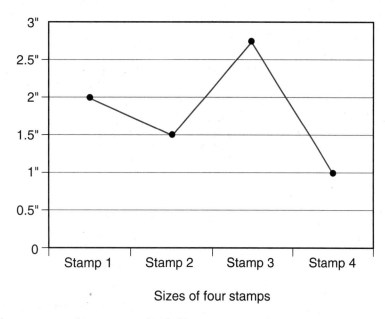

Sizes of four stamps

**8.** The eraser line graph follows:

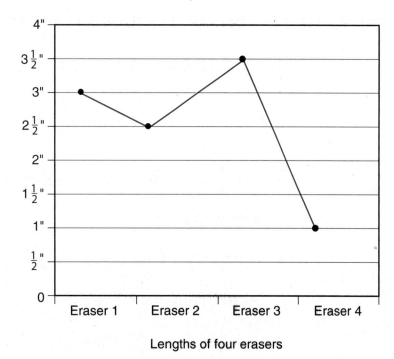

Lengths of four erasers

**9.** The souvenir coins line graph follows:

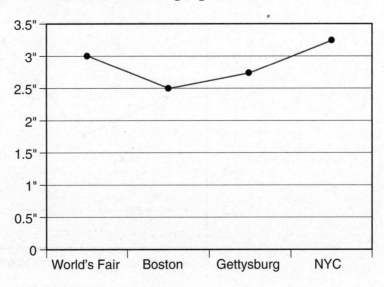

Sizes of four souvenir coins

## ANSWERS TO LIQUID VOLUME MEASURE EXERCISES

**1.** There is $\frac{1}{2}$ liter, or 500 milliliters in the pitcher.

**2.** There is $\frac{1}{2}$ liter, or 500 milliliters in the bottle.

**3.** $\frac{1}{2} + \frac{1}{4} = \frac{3}{4}$. There is $\frac{3}{4}$ liter in the two pitchers combined.

**4.** $1 + 1 + \frac{1}{4} = 2\frac{1}{4}$. There are $2\frac{1}{4}$ liters in the three bottles combined.

**5.** $\frac{3}{4} + \frac{1}{2} + \frac{1}{4} = 1\frac{1}{2}$. There are $1\frac{1}{2}$ liters in the three jars combined.

**6.** $1 + \frac{1}{2} + \frac{3}{4} = 2\frac{1}{4}$. There are $2\frac{1}{4}$ liters in the three bottles combined.

7. $1 - \frac{1}{4} = \frac{3}{4}$, and $\frac{3}{4}$ liter = 750 milliliters. There are 750 milliliters more liquid in the first one than in the second.

8. $1 - \frac{1}{2} - \frac{1}{4} = \frac{1}{4}$. $\frac{1}{4}$ liter = 250 milliliters. There are 250 milliliters more liquid in the first bottle than in the last two.

9. $1\frac{1}{4} \times 5 = 6\frac{1}{4}$. $\frac{1}{4}$ liter = 250 milliliters. There would be 6 liters and 250 milliliters in all if you were to multiply the liquid in the two jars by 5.

10. $\frac{3}{4} \times 6 = 4\frac{1}{2}$. There would be $6\frac{1}{2}$ liters in all if you were to multiply the liquid in the bottle by 6.

11. $2 \div 8 = \frac{1}{4}$. $\frac{1}{4}$ liter = 250 milliliters. Each child would get 250 milliliters of lemonade. That is about a cup (8 ounces) of lemonade.

12. $3 \div 15 = \frac{1}{5}$. $\frac{1}{5}$ liter = 200 milliliters. Each child would get 200 milliliters of iced tea. That is about 6 ounces of iced tea.

## ANSWERS TO VOLUME ESTIMATION EXERCISES

1. You would use a teaspoon to measure 5 teaspoons of liquid.

2. You would use a water bottle to measure 2 liters of a liquid.

3. You would use a bucket to fill a bathtub.

4. You would use a water bottle on a hike to drink water on the trail.

5. You would use a medicine dose cup to measure out 15 milliliters.

6. You would use a liter bottle to measure out 7 liters.

7. You would use a picnic cooler to measure 6 liters.

8. You would use a child's cup to measure 60 milliliters.

## ANSWERS TO REASONING EXERCISES

9. You would use liters to fill up a bucket, because the milliliter is much too small a unit of measure to use when filling up a bucket.

10. You would use milliliters to fill up a small cup, because a liter is much too large a unit to fill up a small cup.

## ANSWERS TO MASS ESTIMATION EXERCISES

1. Some laptop computers are approximately 550 grams. No laptop is close to 5 grams.

2. A pencil is approximately 20 grams.

3. A cell phone is approximately 130 grams.

4. A squirrel is approximately 500 grams.

5. A brick is approximately 2 kilograms.

6. A motorcycle is approximately 50 kilograms.

7. A coffee cup is approximately 140 grams.

8. A shoe is approximately 450 grams.

## ANSWERS TO MASS CALCULATION EXERCISES

1. 300 + 450 = 750. Filomina had 750 grams of sugar when she combined the two amounts.

2. 20 + 12 = 32. Tenille had 32 kilograms of potatoes when she combined the two amounts.

**3.** 450 − 120 = 330. Jorge had 330 grams of coffee left after he used 120 grams.

**4.** 3 − 1.8 = 1.2. Edwina cut 1.2 kilograms off the board.

**5.** 18 − 9 = 9. The pipe was 9 kilograms after Jessica cut 9 kilograms off the pipe.

**6.** Danielle and her six friends make seven people altogether. 56 ÷ 7 = 8. Danielle will measure out 8 grams of rice for her and her friends.

**7.** Walter and his eight friends make nine people altogether. 27 ÷ 9 = 3. Walter measured out 3 grams of cocoa for himself and his friends.

**8.** Tim and his seven friends make eight people altogether. 8 × 1.5 = 12. Tim used 12 grams of tea for himself and his seven friends. That leaves 30 − 12 = 18. 18 grams of tea will be left in the caddy after Tim has made tea for his friends.

**9.** 78 ÷ 6 = 13. Each member of Cosimo's family had 13 grams of salad.

**10.** 2.5 − 1.9 = 0.6. Harry cut 0.6 kilograms off each brick to fit them into the space.

## ANSWERS TO TIME EXERCISES

**1.** C. 1 hour, 12 minutes

**2.** B. 1 hour, 11 minutes

**3.** D. 3:38 P.M.

**4.** A. $3\frac{1}{2} \times 8 = 28$

**5.** C. 9:50 P.M.

## ANSWERS TO TIME (ANALOG CLOCK) EXERCISES

1. 7:20
2. 8:05
3. 9:15
4. 10:12
5. 1:06
6. 3:15
7. 6:12
8. 8:25
9. 5:54
10. 7:31

## ANSWERS TO EXTENDED CONSTRUCTED RESPONSE (MEASUREMENT AND DATA) EXERCISES

1. There are two triangles (the front sails). There are six rectangles (the top two sails on each of the three masts). There are ten squares (the bottom sails on each mast, and the seven cannon ports). There are seven circles (the seven cannons). There is one trapezoid (the sail on the back end, on its own mast).

2. The pyramid has eight edges. The cube has 12 edges. The pyramid has five faces. The cube has six faces.

   a. Similarities: The pyramid and the cube are both approximately the same size. The pyramid and cube are both three-dimensional figures. The pyramid and the cube both have only straight edges. The pyramid and cube both have only flat faces.

   b. Differences: The pyramid and cube each have a different number of faces, and a different number of edges.

## ANSWERS TO DATA ANALYSIS EXERCISES

1. **B.** There are fewer students who like mushrooms than plain.

2. **A.** 12 students were born in the first half of the year.

3. **D.** The SUVs are the most popular vehicles on Gravel Rd.

4.

   **A.** The Walker St. School ordered 20 plain pizzas.

   **B.** The Walker St. School ordered 28 pepperoni pizzas.

   **C.** The Walker St. School ordered eight <u>more</u> plain pizzas than mushroom pizzas.

   **D.** The Walker St. School ordered 12 <u>more</u> pepperoni pizzas than sausage pizzas.

   **E.** The Walker St. School ordered 76 pizzas, altogether.

5.

   **A.** There are 210 pairs of shoes in the ladies shoe department.

   **B.** There are 96 pairs of running shoes and cross-training shoes in the ladies shoe department.

   **C.** There are 18 <u>more</u> pairs of pumps than high heeled shoes in the ladies shoe department.

   **D.** There are 18 <u>fewer</u> pairs of sandals than running shoes in the ladies shoe department.

   **E.** There are 78 <u>fewer</u> pairs of dress shoes than casual shoes in the ladies shoe department.

**6.**

A. Bengal's Bagels sold 60 onion bagels.

B. Bengal's Bagels sold 25 <u>more</u> pumpernickel bagels than plain bagels.

C. Bengal's Bagels sold 25 <u>more</u> onion bagels than whole wheat bagels.

D. Bengal's Bagels sold 65 <u>more</u> onion and sesame bagels than plain bagels.

E. Bengal's Bagels sold 290 bagels in all that cold November day.

**7.**

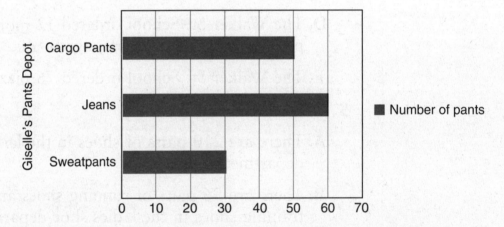

**8.**

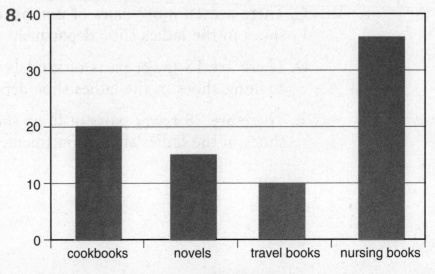

# CHAPTER 6

## ANSWERS TO POLYGON AND QUADRILATERAL EXERCISES

1. **C.** The rectangle is the only polygon of the choices.

2. **D.** Two equilateral triangles will go together, thus:

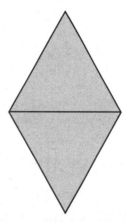

3. **A.** A hexagon. The prefix "hex" means six.

4. **D.** A cone is a three-dimensional figure, not a two-dimensional figure.

5. **A.** 5, pentagon; B. 8, octagon; C. 6, hexagon; D. 3, triangle.

6. The fourth figure is a rhombus.

7. The second figure is a rectangle.

8. The first figure is a trapezoid.

9. The third figure is a square.

10. The third figure is a parallelogram.

## ANSWERS TO LOCATING POINTS ON THE COORDINATE GRID EXERCISES

**1–5.**

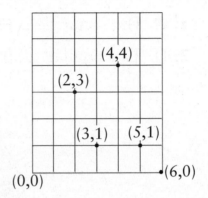

## ANSWERS TO PERIMETER OF SHAPES EXERCISES

1. **A.** The width of the flower bed is 6 feet, and the length of the bed is 13 feet. $6 + 13 + 6 + 13 = 38$ feet. Another way of looking at this: $(6 + 13) \times 2 = 38$ feet.

2. **D.** The box is 6 inches on a side. $6 + 6 + 6 + 6 = 24$ inches. Another way of looking at this is $6 \times 4 = 24$ inches.

3. **C.** The straightaways are each 100 yards, and the end half-rounds are each 120 yards, so the track is $100 + 120 + 100 + 120 = 440$ yards. Another way of looking at this is $(100 + 120) \times 2 = 440$ yards.

4. **B.** The shoebox is 7 inches on one side, 12 inches on another side. $7 + 12 + 7 + 12 = 38$ inches. Another way of looking at it is $(7 + 12) \times 2 = 38$ inches.

5. **D.** The poster is 20 inches on a side, so $20 + 20 + 20 + 20 = 80$. Another way of looking at it is $20 \times 4 = 80$ inches.

6. **C.** Adding the four lengths: $3 + 2 + 5 + 2 = 12$ feet. And so the perimeter is 12 feet.

7. **B.** Each side is 1 foot, so the perimeter is $1 + 1 + 1 + 1 + 1 + 1 = 6$ feet.

## ANSWERS TO AREA OF SHAPES ON A SQUARE GRID EXERCISES

1. The CD case covers 25 square inch blocks, and 5 one-half square inch blocks, so the area is:

   $25 + 2\dfrac{1}{2} = 27\dfrac{1}{2}$ square inches, or $27\dfrac{1}{2}$ in².

2. The book covers 80 square inch blocks, so the area of the book is 80 square inches, or 80 in².

3. The book is 8 inches wide and 10 inches high. The book, therefore, is $8 \times 10 = 80$. The answer is the same as in no. 2.

4. The cereal box covers 70 square inch blocks, and 7 one-half inch square blocks, so the area is:

   $70 + 3\dfrac{1}{2} = 73\dfrac{1}{2}$ square inches, or $73\dfrac{1}{2}$ in².

5. The cereal box is $7\dfrac{1}{2}$ inches wide and 10 inches high. The cereal box, therefore, is $7\dfrac{1}{2} \times 10 = 73\dfrac{1}{2}$ square inches, or $73\dfrac{1}{2}$ in². The answer is the same as in no. 4.

6. The macaroni and cheese box covers 144 square centimeter blocks, so the box is 144 square centimeters, or 144 cm².

7. The macaroni and cheese box is 8 centimeters wide and 18 centimeters high. The macaroni and cheese box, therefore, is $8 \times 18 = 144$ square centimeters, or 144 cm². The answer is the same as in no. 6.

8. The flyer covers 88 square inch blocks, and 8 one-half inch blocks, so the flyer is $88 + 4 = 92$ square inches, or 92 in².

9. The flyer is $8\frac{1}{2}$ inches wide and 11 inches high. The flyer, therefore, is $8\frac{1}{2} \times 11 = 92$ square inches, or 92 in². The answer is the same as in no. 8.

10. The cake box covers 247 square centimeter blocks, and 19 one-half square centimeter blocks, so the box is $247 + 9\frac{1}{2} = 256\frac{1}{2}$ square centimeters, or $256\frac{1}{2}$ cm².

11. The cake box is $13\frac{1}{2}$ centimeters wide and 19 centimeters high. The cake box, therefore, is $13\frac{1}{2} \times 19 = 256\frac{1}{2}$ square centimeters, or $256\frac{1}{2}$ cm². The answer is the same as in no. 10.

12. The cracker box covers 63 square inch blocks, and 9 one-half inch blocks. Therefore, the area is $63 + 4\frac{1}{2} = 67\frac{1}{2}$ square inches, or $67\frac{1}{2}$ in².

13. The cracker box is $7\frac{1}{2}$ inches wide and 9 inches high. The cracker box, therefore, is $7\frac{1}{2} \times 9 = 67\frac{1}{2}$ square inches, or $67\frac{1}{2}$ in². The answer is the same as in no. 12.

14. The cookie box covers 198 square centimeter blocks, and 11 one-half square centimeter blocks, so the box is $198 + 5\frac{1}{2} = 203\frac{1}{2}$ square centimeters, or $203\frac{1}{2}$ cm².

**15.** The cookie box is 11 centimeters wide and $18\frac{1}{2}$

centimeters high. The cookie box, therefore, is

$11 \times 18\frac{1}{2} = 203\frac{1}{2}$ square centimeters, or $203\frac{1}{2}$ cm².
The answer is the same as in no. 14.

**16.** D. 80 ft².

**17.** B. 144 ft².

**18.** A. 63 in².

**19.** C. 28 in².

**20.** A. 5.5 ft².

## ANSWERS TO ODD SHAPES ON A SQUARE GRID

**1.** Courtney's house covers 108 square yards. The patio covers 60 square yards. The porch covers 2 square yards. Therefore, the area of Courtney's house is 108 + 60 + 2 = 170 square yards, or 170 yd².

**2.** The floor Bob is tiling has two floors, a left and a right floor. The left floor covers 20 square meters. The right floor covers 24 square meters. Therefore, the area of the two floors combined is 20 + 24 = 44. The floor is 44 square meters, or 44 m².

**3.** The lawn Andrew is mowing is two areas. The front yard covers 81 square yards, and the side yard covers 52 square yards. Therefore, the area of the 2 yards combined is 81 + 52 = 133 square yards, or 133 yd².

4. The Victorian house covers 169 square yards. The front porch covers 24 square yards. The side porch covers 26 square yards. Therefore, the area of the Victorian house is $169 + 24 + 26 = 219$ square yards, or 219 yd$^2$.

5. The shape Angela made formed three rectangles. The left rectangle covered 30 square centimeters. The middle rectangle covered 36 square centimeters. The right rectangle covered 42 square centimeters. Therefore, the area of the shape Angela made with blocks was $30 + 36 + 42 = 108$ square centimeters, or 108 cm$^2$.

## CHAPTER 7

### ANSWERS TO PROBLEM-SOLVING EXERCISES

1. Greg is planning his barbeque. He needs to decide when to put the various items on his grill. It would seem that if we worked backward from 3:00 P.M., we'll get our answer.

| Hamburgers | Ribs | Hot Dogs | Chicken |
|---|---|---|---|
| 3:00 p.m. <br> - 20 min <br> 2:40 p.m. | 3:00 p.m. <br> - 35 min <br> 2:25 p.m. | 3:00 p.m. <br> - 10 min <br> 2:50 p.m. | 3:00 p.m. <br> - 30 min <br> 2:30 p.m. |

So, if he starts at 2:25 P.M. putting the ribs on, then at 2:30 P.M. he puts the chicken on, at 2:40 P.M. he puts the hamburgers on, and finally at 2:50 P.M. he puts the hot dogs on, he'll be able to take all of them off at 3:00 P.M.

**2.** Childrens' admission price is $7 and the adults' price is $12, so some combination of sevens and twelves that add up to 104 should give us our answer. One restriction we need to remember is that there were more children than there were adults. Guess and check appears to be a good way to solve this. Let's try seven children and six adults:

$$7 \times 7 = 49$$
$$12 \times 6 = \underline{72}$$
$$\$121$$

That's too much. But, we see that we'll get an odd number if we multiply 7 by an odd number. We need to multiply 7 by an even number to get 104. Let's try eight children and five adults:

$$7 \times 8 = 56$$
$$12 \times 5 = \underline{60}$$
$$\$116$$

Again, that's too much. But, it's too much by just $12. Let's drop one of the adults:

$$7 \times 8 = 56$$
$$12 \times 4 = \underline{48}$$
$$\$104$$

We've got it. Do you see how, in each case, the previous guess helped to narrow choices for the next guess, and helped us find the answer? Always try to think of how the last guess helps you toward the next guess. This is one of the most powerful aspects of the guess and check method: its ability to narrow our choices, getting us closer to the answer with each guess.

**3.** We need to find out how far they have kayaked and how far they are from their starting point. Drawing a picture would seem to work for this one, but not before we calculate how far they have paddled one way and the other.

They paddled eastbound from 8:00 A.M. till noon. That is 4 hours. They went 5 mph during that time. 4 · 5 = 20 mi. They traveled 20 miles east. Then, they traveled from 1:00 P.M. till 5:00 P.M. That is 4 hours. They traveled an a speed of 4 mph during that time. 4 · 4 = 16 mi. They traveled 16 miles west. So, let's draw a picture of this:

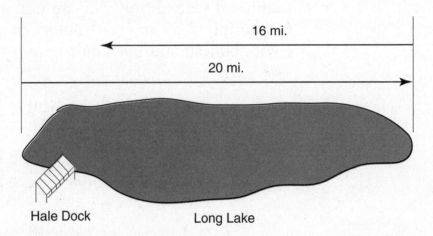

Hale Dock          Long Lake

As you can see, Phil and Ben are not at their original starting place. To find out how short they are, subtract 16 from 20: 20 − 16 = 4. They are 4 miles short of their starting point. This is a good example of a problem that can be solved by a combination of a picture (or drawing) and a few simple equations.

**4.** Andrew picks up 6 bottles/cans for every 5 bottles/cans Henry picks up, and we know that the total picked up is 132. We want to know how many each boy picked up. If we made up a table, we could keep track of how much each boy picked up and of the running total.

| Andrew | Henry | Total |
|--------|-------|-------|
| 6 | 5 | 11 |
| 12 | 10 | 22 |
| 18 | 15 | 33 |
| 24 | 20 | 44 |
| 30 | 25 | 55 |
| 36 | 30 | 66 |
| 42 | 35 | 77 |
| 48 | 40 | 88 |
| 54 | 45 | 99 |
| 60 | 50 | 110 |
| 66 | 55 | 121 |
| 72 | 60 | 132 |

So, Andrew picked up 72 bottles/cans, and Henry picked up 60 bottles/cans. This gives us a total of 132 bottles/cans.

**5.** We need to find out how many dots are in an array that has 10 on a side. A chart, along with a picture, would help us to see the solution to this.

As you can see, a triangle with one dot has one dot in it, an array with two dots on a side has three dots in it, and an array of dots having three dots on a side has 6 dots in it. Let's look at this in a table:

| Triangle | Dots |
|----------|------|
| 1        | 1    |
| 2        | 3    |
| 3        | 6    |
| 4        | 10   |
| 5        | 15   |
| 6        | 21   |
| 7        | 28   |
| 8        | 36   |
| 9        | 45   |
| 10       | 55   |

So, the answer is that a triangular array of dots that has 10 dots on a side will contain 55 dots.

## ANSWERS TO PROBLEM SOLVING AND OTHER MATH SKILLS

For all of these problems we'll use the four-step process.

1. a) **Make sense of the problem:** We're trying to find the beginning number in a sequence of operations. What are the facts? We have the sequence of numbers, and the operations, but not the beginning number. We know all the operations and all the numbers, except the beginning one. How can we do this?

b) **Make a plan:** Can we draw a picture? That won't help. Can we look for a pattern? Not really, because there is no pattern to see. Can we work backward? Well, that might help, but we need to know what we started with. Can we guess and check? Yes, that seems to be the right way to proceed.

c) **Carry out the plan:** So, let's try a starting number. The number should be odd, because we add 5 right away, and then divide by 2, so we need an even number. Starting with an odd number will insure an even number when we have to divide by 2. Let's try 9:

$9 + 5 = 14 \div 2 = 7 - 4 = 3 + 9 = 12 \cdot 2 = 24 \div 5 =$ Oh, oh. We can't divide 24 by 5 evenly, and our final answer is 8. We know that 9 does not work, but we need a number larger than that number. Let's try 15: $15 + 5 = 20 \div 2 = 10 - 4 = 6 + 9 = 15 \cdot 2 = 30 \div 5 = 6$. We're not far away from 8 (just two off). But, by using a multiple of 5, we seemed to get a whole number for an answer. Let's try 25: $25 + 5 = 30 \div 2 = 15 - 4 = 11 + 9 = 20 \cdot 2 = 40 \div 5 = 8!$ We've found it. Did you see that we gleaned clues as we went along for how to "choose smarter" on the next pick? Try to do that all the time.

d) **Check your answer:** does this work? Yes, because when you plug the number in you'll get 8. Does this make sense? Yes, it does. Are there other numbers that would work? Probably not. So, this works.

2. a) **Make sense of the problem:** We're trying to find the perimeter to get the proper amount of fence. The garden is shaped like a trapezoid. How will we do this?

b) **Make a plan:** This is not a difficult problem. If we get the perimeter, we'll know the amount of fence we need. We'll add up all the legs of the trapezoid. It might be helpful to draw a diagram of the garden.

c) **Carry out the plan:** We have $4 + 6 + 5 + 8 = 23$. That's 23 feet. When you know units, you should put them in.

d) **Check your answer:** Does this make sense? Yes it's reasonable that 23 feet of fence will adequately fence in this garden. Are there other answers that will work? Not really, because we need an exact amount, here. So this works.

3. a) **Make sense of the problem:** We need to know how many numbers between 100 and 200 have a 6 in them. What are the facts? The numbers from 100–200. How will we do this?

b) **Make a plan:** Can we guess? That won't work, because we'll have no good way to check it, except to look at all the numbers, which will take a long time... Can we make a picture? A picture won't solve the problem. Can we look for a pattern? Yes, that probably will get the answer for us. So let's try a pattern.

c) **Carry out the plan:** How will we set up the plan? Let's look at the first ten numbers. 100, 101, 102, 103, 104, 105, 106, 107, 108, 109, 110. So, for the first ten numbers, a 6 appears once. So, for each group of ten numbers, a 6 will appear once. Is this true? Let's check the next ten numbers: 111, 112, 113, 114, 115, 116, 117, 118, 119, 120. Yes, that works. There are ten groups of these tens, so in the ten groups of tens from 100–200, there are ten numbers that have a 6 in them. Is that right? Oh, wait! When we get to the numbers

160 to 169, all ten numbers have a 6 in them. OK, so there are ten more numbers: $10 + 10 = 20$. Is that right? Well, not really, because one of those numbers, 166, is being counted twice, so we need to take one away from that 20. There are 19 numbers between 100 and 200 that have a 6 in them.

d) **Check your answer:** Does this make sense? Well, we looked at the numbers, found a pattern, and followed that pattern. It's reasonable that there are 19 numbers that have a 6 in them.

4. a) **Make sense of the problem:** What is the required answer? We want the last number in the sequence. What are the facts? We have the sequence of numbers. How can we get this? Figure out the pattern (how we get from one number to the next) and then apply that rule to the last number to get the required number.

b) **Make a plan:** How will we solve this? Searching for a pattern seems to be the best way to approach this problem. We'll examine each number, and see how to get from one number to the next, and see what the pattern is to do that. Then we'll try to apply that pattern to the next number in the sequence.

c) **Carry out the plan:** Let's look at all the numbers in the sequence:

$$23 \quad 28 \quad 26 \quad 31 \quad 29 \quad 34\ldots$$

This is a very strange sequence. It goes up, then down, then up again. What number does it go up by? 5, initially, then down 2, then up 5, then down 2, then up 5 again. So, the next number in the sequence would be down by 2 or 32.

d) **Check your answer:** Does this sound reasonable? Yes, it does. The pattern is: 5 up, 2 back, and repeat. So following that pattern, the next number will be the last number minus 2, or 32.

**5.** a) **Make sense of the problem:** What is the required
answer? We want to find out the distance from each
side of the front yard to plant a tree. What are the
facts? We have the diameter of the dirt ball, and the
width of the yard. How can we get this? We can draw
a picture to find out where to plant the dirt ball of the
tree.

b) **Make a plan:** We'll draw a picture so that we can
figure how far from each side Peter needs to put the
dirt ball so that it is in the center of the yard. Peter
might also look at how to center it using only numbers.

c) **Carry out the plan:**

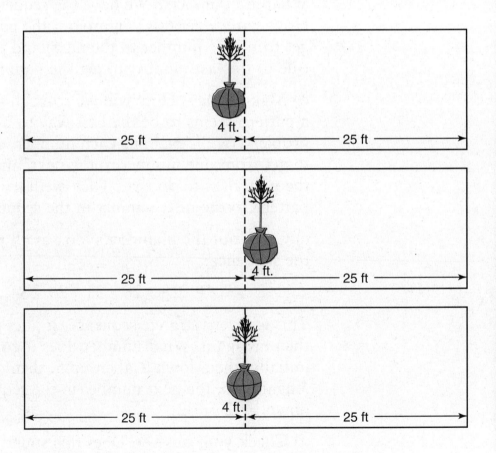

You see that if Peter puts the dirt ball at 25 ft., it
would extend 4 ft. on one side or the other, so it
wouldn't be centered. He must place it so that 2 ft.

are extended on each side of the center, so that it is centered.

There is another way. The yard is 50 ft. long, and half of that is 25. The dirt ball is 4 ft., so cut it in half, giving 2 ft. on each side. So the dirt ball will start at 23 ft. and end at 27 ft. on the number line, and that will center it.

d) **Check your answer:** Does this answer the question? Yes, it does. It places the dirt ball exactly in the center of the yard. Peter saw two ways of doing it here, and many problems can be done in more than one way.

6. a) **Make sense of the problem:** What is the question being asked? How many different sums of money can be made with four coins? What are the facts? Mickey has 2 pennies, a nickel, and a quarter. Note that using one or the other penny will not give another sum of money. Mickey wants only to find different sums.

b) **Make a plan:** Will a picture help? Not really. A chart probably will be best, because it will show all the possibilities very clearly. So Mickey will make a chart.

c) **Carry out the plan:** Here's the chart

| 1 cent | 1 | 2 | 1 | 2 | 1 | 2 |  | 1 | 2 |
|---|---|---|---|---|---|---|---|---|---|
| 5 cents |  |  | 1 | 1 |  |  | 1 | 1 | 1 |
| 25 cents |  |  |  |  | 1 | 1 | 1 | 1 | 1 |
| sum | 1 | 2 | 6 | 7 | 26 | 27 | 30 | 31 | 32 |

So we have 9 sums.

d) **Check your answer:** Does this give all the sums? Yes, it does. Mickey looked at all the possible combinations from one of each coin to all four coins together. All the possible sums are here.

**7.** a) **Make sense of the problem:** What is the desired quantity? How many free tickets will be given? What are the facts? We know that for every 8 tickets, we get one free, and we need 50 tickets. How will we do this?

b) **Make a plan:** If we take groups of 8 away from the school group, and then take away one more (to account for the free ticket, 9 in all), we'll find out how many free tickets we will get.

c) **Carry out the plan:** We start with 50:

$$50 - 9 = 41$$
$$41 - 9 = 32$$
$$32 - 9 = 23$$
$$23 - 9 = 14$$
$$14 - 9 = 5$$

5 – 9 ? We cannot take 9 from 5, so we're done. We don't have another 8 students to get a free ticket, so we'll get 5 free tickets.

d) **Check your answer:** Does this answer the question? Yes, it does. See how repeated subtraction will get us to our answer.

**8.** a) **Make sense of the problem:** What are we to find? We need to name the shape formed by the four points, and we need to find the perimeter of the shape. Then, we need to slide the figure up four units, and calculate where the points would be that form its corners.

b) **Make a plan:** The best way to solve this is to plot the points on a graph and connect them. Then, we can see what shape is made.

c) Carry out the plan:

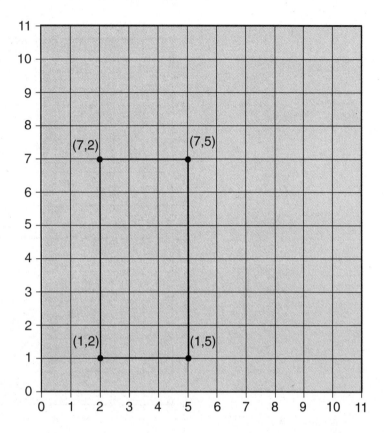

A. We see that the figure made is a rectangle.

B. We see that the rectangle's width is 3 units, and the length is 6 units, so the perimeter is $3 + 6 + 3 + 6 = 18$.

C. If we slide the rectangle up four units, we get this

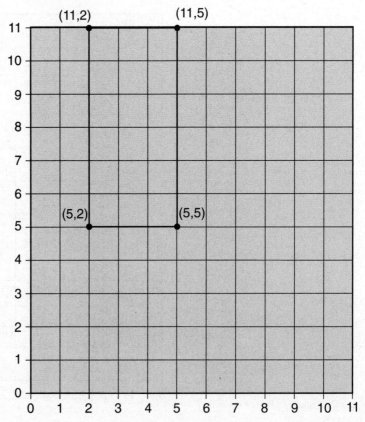

d) **Check your answer:** Does this answer the questions? Yes, it does. Does it seem reasonable? Yes.

9. a) **Make sense of the problem:** What is the required answer? It's to know how many books the bookstore sold on the first day, what day the sales went down, and how many books were sold for all six days we are looking at. What are the facts? We have a pictograph where each book represents 10 books. How will we do this?

b) **Make a plan:** It would seem that a chart, or table, will show the amounts sold more clearly. A chart is what we should use. We should expand the number of books out, however, giving the proper number of books for each.

c) **Carry out the plan:** Here is the chart:

| Day | Books Sold |
|-----|-----------|
| 1 | 30 |
| 2 | 45 |
| 3 | 50 |
| 4 | 60 |
| 5 | 55 |
| 6 | 70 |

So, the answers to the three questions are:

    A.  Bob and Ray sold 60 books on the fourth day.
    B.  Sales went down on the fifth day.
    C.  The total amount of books sold in the six days was:

$$30 + 45 + 50 + 60 + 55 + 70 = 310 \text{ books}$$

d) **Check your answer:** Does this answer all the questions? Yes, it does. It's much easier to read when you put the information in a table.

10. a) **Make sense of the problem:** What is the required answer? How many different combinations of outfits can Martina make? What are the facts? The number of blouses Martina has (4), and the number of pants she has (3). How will we do this?

b) **Make a plan:** The way we can solve this is to make a tree diagram.

c) **Carry out the plan:** Martina will make a diagram that matches up all the blouses with all the pants.

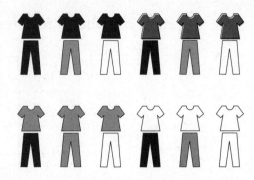

You see that there are 12 combinations of blouses with pants.

If she had two pairs of shoes, how many combinations of blouses, pants, and shoes will Martina have?

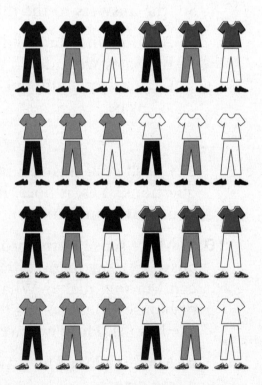

You see that there are 24 combinations of blouses, pants, and shoes.

d) **Check your answer:** Does this make sense? Yes, it does. You can see clearly, with a tree diagram, *all* the combinations that are available to Martina when she goes on vacation.

# SAMPLE TESTS AND ANSWERS

## PRACTICE TEST 1

### DAY 1

### Part 1

---

### Directions

This part of the test contains eight short constructed response questions. You may *not* use a calculator for these questions. Write your answers on the line.

---

1.  Add 362 + 537.

    Write your answer here __899__

    $$\begin{array}{r} +362 \\ +537 \\ \hline 899 \end{array}$$

2.  Round 687 to the nearest tens.

    Write your answer here __690__

3.  Daphne asked her classmates in school to help on a food drive. On Monday, she collected five items of food. On Tuesday, she collected 12 items. On Wednesday, she collected 20 items. On Thursday, she collected 10 items. On Friday, she collected six items. How much did Daphne collect that week?

    Write your answer here __53 items__

    $$\begin{array}{r} 5 \\ 12 \\ 20 \\ 10 \\ +6 \\ \hline 53 \end{array}$$

**Go On**

4. Calculate the value of *h* when
   *h* − 23 = 39.

   Place your answer here ___62 ✓___

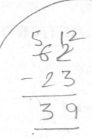

$$\begin{array}{r} 5\;\;12 \\ \cancel{6}\;\cancel{2} \\ -\;2\;3 \\ \hline 3\;9 \end{array}$$

5. Regina keeps track of the customers
   coming to her car dealership. In the
   months of summer, she tracked these
   numbers of customers:

   | Month | Customers |
   | --- | --- |
   | May | 16 |
   | June | 23 |
   | July | 31 |
   | August | 25 |

   How many customers came to her
   dealership in the summer?

   Place your answer here ___90 customers___

$$\begin{array}{r} +\;1\,6 \\ 2\,3 \\ \hline +\;3\,9 \\ 3\,1 \\ \hline 7\,0 \\ +\;2\,5 \\ \hline 9\,5 \end{array}$$

6. The following pictograph shows the number of pizzas made at Althea's Pizza Place on the day of the Super Bowl.

Key: = 10 pizzas

How many pizzas did Althea's Pizza Place make on the day of the Super Bowl?

Place your answer here ___90___

7. Ben works at a carwash drying cars. Last week he dried the following number of cars:

| Day | Monday | Tuesday | Wednesday | Thursday | Friday |
|---|---|---|---|---|---|
| Cars dried | 9 | 18 | 27 | 36 | |

If this pattern continues, how many cars will he dry on Friday?

Place your answer here ___45___

**Go On**

**8.** Find the value of *g* in this equation:
   $56 = 8 \times g$

STOP

If you have time, you may review your work in this section only.

**PART 2**

---

## Directions

This part of the test has 11 multiple-choice questions. You may *not* use a calculator for these questions. Fill in the circle next to your choice.

---

9. Billie is washing windows at her neighbor's house. She washed 22 windows on the first floor; then she went to the second floor and washed 18 windows. Finally, she went to the third floor and washed 12 windows. How many windows did she wash in all?

    Ⓐ 28
    Ⓑ 30
    Ⓒ 54
    Ⓓ 52

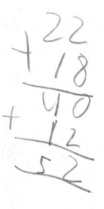

**Go On**

**10.** Ahmed went to the store to buy some peppers. They come in packages of four:

| Peppers per package | Packages of peppers bought | Total number of peppers |
|---|---|---|
| 4 | 6 | ? |

If Ahmed bought six packages, how many peppers did he buy?

Ⓐ 18

Ⓑ 24

Ⓒ 28

Ⓓ 30

**11.** Find the exact answer for 380 + 170.

Ⓐ 550

Ⓑ 490

Ⓒ 660

Ⓓ 530

**12.** In the following fact triangle, what is the value of *r*?

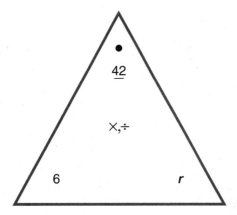

Ⓐ 4

Ⓑ 5

Ⓒ 6

Ⓓ 7

**13.**

Hilda noticed the clock above. What time did she read?

Ⓐ 6:40

Ⓑ 6:50

Ⓒ 6:45

Ⓓ 6:35

**Go On**

**14.** George wanted to make some rice. He looked on the box and read this:

| Rice (cups) | 1 | 2 | 3 |
|---|---|---|---|
| Water (cups) | $1\frac{1}{2}$ | 3 | $4\frac{1}{2}$ |

If the pattern continues, how much water will he need for 4 cups of rice?

Ⓐ  5 cups water

Ⓑ  $5\frac{1}{2}$ cups water

Ⓒ  6 cups water

Ⓓ  $6\frac{1}{2}$ cups water

**15.** Tatiana had 234 stamps in her collection. Dan gave her some more, and then she had 458 stamps. How many stamps did Dan give her?

$$\begin{array}{r} 458 \\ -\ 234 \\ \hline 224 \end{array}$$

Ⓐ  230

Ⓑ  224

Ⓒ  222

Ⓓ  220

16. Harry had a jar filled with 864 pennies. What is the value of the 6 in that number?

   Ⓐ 6 thousands

   Ⓑ 6 ones

   Ⓒ 6 tens

   Ⓓ 6 hundreds

17. What is the name of a four-sided polygon where all the sides are the same length, and all the angles are 90 degrees?

   Ⓐ A rectangle

   Ⓑ A trapezoid

   Ⓒ A square

   Ⓓ A rhombus

**Go On**

18. Alvin has a collection of toy boats. He
has 32, in all, and displays them in a
book case that has four shelves,
according to the following table:

| Boats on each shelf | Boats on each shelf | Total number of boats |
|---|---|---|
| 4 | ? | 32 |

If he stores an equal number of boats
on each shelf, how many will be on
each shelf?

Ⓐ  8 boats

Ⓑ  9 boats

Ⓒ  10 boats

Ⓓ  11 boats

19. Which symbol goes between the two
    fractions shown?

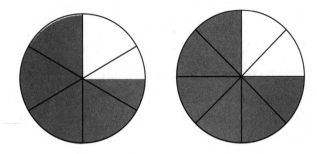

Ⓐ  >

Ⓑ  =

Ⓒ  <

Ⓓ  None of the above

If you have time, you may review
your work in this section only.

**PART 3**

---

### Directions

This part of the test contains eight multiple-choice questions and one extended constructed response question. You may *not* use a calculator for these questions. Fill in the circle next to your choice. Follow the directions for the extended constructed response question.

---

20.   Find the exact answer for 844 − 567.

    Ⓐ 267

    Ⓑ 281

    Ⓒ 277

    Ⓓ 254

**1**

21.

Juan noticed the clock above. What time did he read?

(A) 9:20

(B) 9:23

(C) 9:35

(D) 9:28

22. Fredrica has a trading card collection. She has 67 baseball cards and 53 hockey cards. Which of the following expressions could you use to describe how many cards she has in all?

(A) 67 + _____ = 53

(B) _____ − 53 = 67

(C) 67 − 53 = _____

(D) 67 + 53 = _____

**Go On**

23. Use your ruler to answer the following question. How long is the following line?

———————————————

Ⓐ  $2\frac{1}{2}$ inches

Ⓑ  $3\frac{1}{4}$ inches

Ⓒ  $2\frac{3}{4}$ inches

Ⓓ  $2\frac{1}{4}$ inches

24.

In the above figure, how much of the chocolate bar remains from the original?

Ⓐ  $\frac{2}{3}$

Ⓑ  $\frac{1}{3}$

Ⓒ  $\frac{5}{6}$

Ⓓ  $\frac{1}{2}$

**25.** In the following fact triangle, what is the value of *s*?

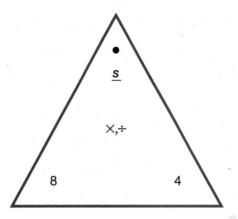

(A) 28

(B) 32

(C) 36

(D) 40

**26.** Hans works at a restaurant, and he is slicing onions. He can slice three onions in one minute. At that rate, how long will it take for him to slice 21 onions?

(A) 7 minutes

(B) 3 minutes

(C) 6 minutes

(D) 9 minutes

**Go On**

27. Which ordered pair shows where
    Shelly's house is?

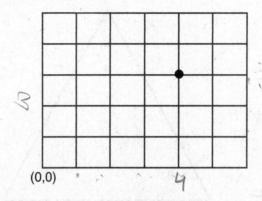

Ⓐ  (2, 3)

Ⓑ  (3, 4)

Ⓒ  (4, 5)

Ⓓ  (4, 3)

## DIRECTIONS FOR THE EXTENDED CONSTRUCTED RESPONSE QUESTION

This question is an extended response question. Remember to:

- Read the problem and think about the answer
- Answer all parts of the question
- Show all your work
- A drawing may be used to show your answer

28. Earl has a small shop selling newspapers. One month, he sold papers according to the following chart:

| Week | Newspapers |
|------|------------|
| 1 | 170 |
| 2 | 340 |
| 3 | 510 |
| 4 | |
| 5 | |
| 6 | |

*(handwritten in table: + 170, + 680, + 850, 1,020)*

*(handwritten on left:)*
I knew 170 was the papers in one month. In two months, it's 170+ 170, so I counted the pattern to my answer.

If the pattern continues, how many papers will have been sold by week 6? Please show all your work or explain your answer.

*(handwritten:)* answer: 1,020  stop

**STOP**

If you have time, you may review your work in this section only.

*(handwritten:)* stop

## DAY 2

**PART 4**

### Directions

This part of the test contains eight multiple-choice questions and one extended constructed response question. You *may* use a calculator for these questions. Fill in the circle next to your choice. Follow the directions for the extended constructed response question.

29.    Find the perimeter of this rectangle:

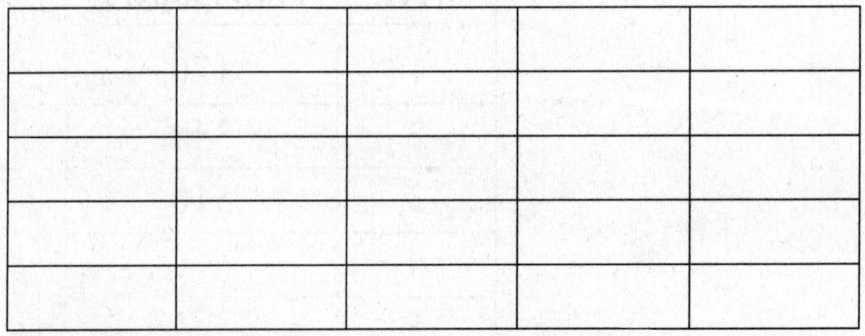

Ⓐ  15

Ⓑ  20

Ⓒ  25

Ⓓ  30

30. What sign should come between these two fractions?

$$\frac{1}{2} \; ? \; \frac{3}{8}$$

Ⓐ >

Ⓑ <

Ⓒ =

Ⓓ None of the above

31. In the following fact triangle, what is the value of *d*?

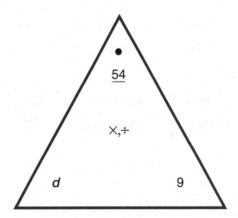

Ⓐ 7

Ⓑ 6

Ⓒ 8

Ⓓ 9

**Go On**

32. Jamie has a collection of hip-hop CDs. He had 237, and then he sold some to his friend Hattie, and then he had 189. What number sentence could you use to show the number of CDs he sold to Hattie?

Ⓐ  $237 + \underline{\hspace{1.5cm}} = 189$

Ⓑ  $237 + 189 = \underline{\hspace{1.5cm}}$

Ⓒ  $189 - 237 = \underline{\hspace{1.5cm}}$

●  $237 - \underline{\hspace{1.5cm}} = 189$

33. Kim is planning her birthday party and wants to put 10 sour candies in the goodie bag for each of her eight friends according to the following table:

| Bags of candies | Candies per bag | Total number of candies |
|-----------------|-----------------|-------------------------|
| 9               | 10              | ?                       |

If she prepares a bag for all her friends and one for herself, how many sour candies will she need?

Ⓐ  80

Ⓑ  70

●  90

Ⓓ  100

**34.** In this IN–OUT table,

| RULE |
| :---: |
| ×3 |

| IN | OUT |
| :---: | :---: |
| 13 | 39 |
| 5 | 15 |
| ? | 24 |

what number goes in the ?

(A) 5

(B) 6

(C) 7

(D) 8

**35.** Simon's class is going on a trip to the aquarium. There are 54 students in the class. The buses that can take the class will each hold nine students. How many buses will be needed to bring the class to the aquarium?

(A) 6 buses

(B) 5 buses

(C) 7 buses

(D) 4 buses

**Go On**

36.

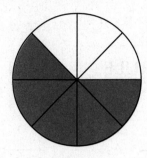

In the figure above, what is the amount that is gone (the amount missing)?

(A) $\dfrac{6}{8}$

(B) $\dfrac{3}{4}$

(C) $\dfrac{3}{6}$

(D) $\dfrac{3}{8}$

## DIRECTIONS FOR THE EXTENDED CONSTRUCTED RESPONSE QUESTION

This question is an extended response question. Remember to:

- Read the problem and think about the answer
- Answer all parts of the question
- Show all your work
- A drawing may be used to show your answer

37. Hillary's class has more boys than girls. Her teacher lines the students up. She is able to put three boys, then one girl, then three boys, and so on, until the whole class is lined up. If there are 28 students in the class, how many are boys, and how many girls are in the class?

7 girls and 22 boys.

It you have time, you may review your work in this section only.

## PART 5

---

### Directions

This part of the test contains eight multiple-choice questions and one extended constructed response question. You may *not* use a calculator for these questions. Fill in the circle next to your choice. Follow the directions for the extended constructed response question.

---

38. Ginger went on a trip with her family. They flew from Newark to Baltimore in 1 hour. Then they flew to Cincinnati from Baltimore in $2\frac{1}{2}$ hours. From Cincinnati, they flew to Dallas in 3 hours. Finally, they flew from Dallas to Denver in 4 hours. How many hours, in total, were they in the air?

    Ⓐ 10 hours

    Ⓑ 9 hours

    Ⓒ $10\frac{1}{2}$ hours

    Ⓓ 11 hours

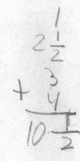

39. Find the exact answer to 724 − 336.

    Ⓐ  338

    Ⓑ  358

    Ⓒ  378

    Ⓓ  388

40. In the following fact triangle, what is the value of *h*?

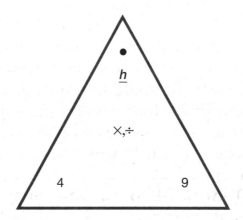

    Ⓐ  36

    Ⓑ  35

    Ⓒ  32

    Ⓓ  28

**Go On**

41. Yuri is an Olympic track and field champion. He has won five gold medals, 18 silver medals, and 29 bronze medals in his career. How many medals did he win in all?

Ⓐ 46

Ⓑ 49

Ⓒ 52

Ⓓ 55

42. For his birthday, Hannon wants to bake cookies for his class. His class has 23 students in it and he wants to bake enough for four cookies per student. How many cookies will he need to bake?

Ⓐ 86

Ⓑ 92

Ⓒ 88

Ⓓ 96

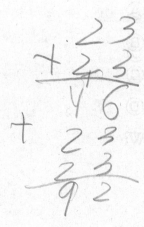

**43.** What is the area of this rectangle? The squares are centimeters.

|  |  |  |  |  |  |  |  |
|--|--|--|--|--|--|--|--|
|  |  |  |  |  |  |  |  |

Ⓐ 16 square centimeters

Ⓑ 18 square centimeters

Ⓒ 20 square centimeters

Ⓓ 22 square centimeters

**44.** Lenny's second grade class exchanged valentines with the third grade class. There are 24 students in Lenny's class, and 31 students in the third grade. How many more students are in the third grade?

Ⓐ 7

Ⓑ 9

Ⓒ 11

Ⓓ 13

**45.** What is $e$ in the equation $5 \times e = 70$?

Ⓐ 350

Ⓑ 250

Ⓒ 14

Ⓓ 15

**Go On**

## DIRECTIONS FOR THE EXTENDED CONSTRUCTED RESPONSE QUESTION

This question is an extended response question.
Remember to:

■ Read the problem and think about the answer
■ Answer all parts of the question
■ Show all your work
■ A drawing may be used to show your answer

46.  Inga drove from Boston to Newark.
     The trip is 324 miles. She gets 27 miles
     for each gallon of gasoline she uses.
     How many gallons will she use to make
     the trip from Boston to Newark?

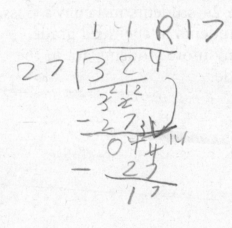

If you have time, you may review
your work in this section only.

## PART 6

### Directions

This part of the test contains eight multiple-choice questions and one extended constructed response question. You may *not* use a calculator for these questions. Fill in the circle next to your choice. Follow the directions for the extended constructed response question.

47. Kayla was examining the following IN–OUT table:

| RULE |
| :---: |
| ×? |

| IN | OUT |
| :---: | :---: |
| 6 | 24 |
| 7 | 28 |
| 4 | 16 |

What rule does Kayla observe?

Ⓐ ×5

**Ⓑ** ×4

Ⓒ ×3

Ⓓ ×6

**Go On**

48. Manfred has a collection of toy cars and trucks. He has 73 cars and 38 trucks. How many more cars does he have than trucks?

    Ⓐ 35

    Ⓑ 45

    Ⓒ 25

    Ⓓ 55

49. Find the exact answer to 453 + 378.

    Ⓐ 831

    Ⓑ 731

    Ⓒ 821

    Ⓓ 721

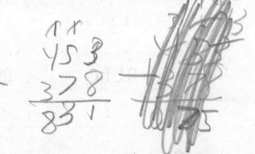

50. Nicole has 250 Spiderman stickers and 92 Captain America stickers. How many <u>more</u> Spiderman stickers does she have?

    Ⓐ 148

    Ⓑ 168

    Ⓒ 162

    Ⓓ 158

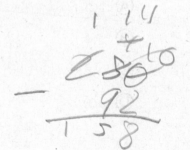

**51.**

In the figure above, how much liquid is in the pitcher?

Ⓐ $\frac{1}{2}$ liter

Ⓑ $\frac{3}{4}$ liter

Ⓒ $\frac{1}{4}$ liter

Ⓓ None of the above

**52.** Find the exact answer to 60 ÷ 6.

Ⓐ 13

Ⓑ 12

Ⓒ 20

Ⓓ 10

**Go On**

**53.** Erin was on a hike with her girl scout troop. Her troop was 14 girl scouts, including Erin, and they had 21 liters of water among the troop. How much water would each girl scout get, if they are dividing it up evenly?

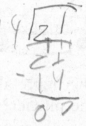

Ⓐ 1 liter

Ⓑ 2 liters

Ⓒ 1.5 liters

Ⓓ 2.5 liters

**54.** What symbol should go between these two fractions?

$$\frac{3}{4} \underline{\hspace{1.5cm}} \frac{5}{8}$$

Ⓐ >

Ⓑ <

Ⓒ =

Ⓓ None of the above

## DIRECTIONS FOR THE EXTENDED CONSTRUCTED RESPONSE QUESTION

This question is an extended response question. Remember to:

■ Read the problem and think about the answer
■ Answer all parts of the question
■ Show all your work
■ A drawing may be used to show your answer

**55.** Halstead & Company makes solar panels to help homeowners get off the electrical grid. They decided to go to *Green NJ*, an all-day environmental fair. They brought 200 solar calculators to give away. They also got names of people to sign up for a consultation on solar panels for their homes. In the morning, they gave away 86 calculators, and for every three calculators they gave away, they got one person to sign up. In the afternoon, they gave away 45 calculators, and they signed up one person for every three calculators they gave away. How many calculators did they have left? How many people did they sign up for consultations?

*(handwritten work)*

$$
\begin{array}{r}
5 1 \\
+ \; 3 1 \\
\hline
2 6 2
\end{array}
$$

$$
\begin{array}{r}
2 0 0 \\
- 1 3 1 \\
\hline
0 6 9
\end{array}
$$

Had 69 calculators left.

$$
\begin{array}{r}
+ \quad 45 \\
86 \\
\hline
1 3 1
\end{array}
$$

$$
\begin{array}{r}
1 \; R 6 9 \\
1 3 1 \overline{\smash{)}2 0 0} \\
- 2 0 0 \\
- 1 3 1 \\
\hline
0 6 9
\end{array}
$$

**STOP**

If you have time, you may review your work in this section only.

# ANSWERS TO PRACTICE TEST 1

**DAY 1**

**PART 1**

1. $362 + 537 = 899$.

2. 687 rounded to the nearest tens is 690.

3. Daphne and her classmates collected 5, 12, 20, 10, and 6 items. So, they collected: $5 + 12 + 20 + 10 + 6 = 53$ items for the food drive.

4. $h - 23 = 39 \rightarrow h - 23 + 23 = 39 + 23 \rightarrow h = 62$.

5. Regina counted 16, 23, 31, and 25 customers in her car dealership. That's $16 + 23 + 31 + 25 = 95$. There were 95 customers who came into Regina's car dealership during the summer.

6. Althea's Pizza Place made $9 \times 10 = 90$ pizzas for the Super Bowl that year.

7. If the pattern continues, Ben will have dried 45 cars on Friday.

8. $56 = 8 \times g \rightarrow \dfrac{56}{8} = \dfrac{8}{8} \times g \rightarrow 7 = g$.

**PART 2**

9. D. 52. $22 + 18 + 12 = 52$.

10. B. 24. $4 \times 6 = 24$.

11. A. 550. $380 + 170 = 550$.

12. D. 7. $7 \times 6 = 42$.

13. C. 6:45. The time on the clock is 6:45.

14. C. 6. George will need 6 cups of water, if the pattern continues.

15. B. 224. $234 + 224 = 458$.

16. **C.** tens. In 864, the 6 is the numeral in the tens place.

17. **C.** A square. A four-sided polygon in which all the sides are the same size, and where all the angles are 90° is a square.

18. **A.** eight boats. $4 \times 8 = 32$.

19. **B.** =. The fractions are the same size.

## PART 3

20. **C.** 277. $844 - 567 = 277$.

21. **B.** 9:23. The time on the clock is 9:23.

22. **D.** $67 + 53 = $ _____.

23. **C.** $2\frac{3}{4}$ inches. The line is $2\frac{3}{4}$ inches long.

24. **A.** $\frac{2}{3}$. There are eight of the original 12 pieces of the bar remaining, and that reduces to $\frac{2}{3}$.

25. **B.** 32. $8 \times 4 = 32$.

26. **A.** 7 minutes. $3 \times 7 = 21$.

27. **B.** (3, 4).

## EXTENDED CONSTRUCTED RESPONSE QUESTION

28. In this question, we are looking for the extension of the pattern that is established in the first 3 weeks. Observing the first 3 weeks, the pattern is that each week, there will be 170 more papers than the previous week. Thus, the chart (completely filled in) is:

| Week | Paper |
|------|-------|
| 1 | 170 |
| 2 | 340 |
| 3 | 510 |
| 4 | 680 |
| 5 | 850 |
| 6 | 1020 |

The number of papers Earl sold in week 6 then, if the pattern continues, is 1,020.

**DAY 2**

**PART 4**

29. **B.** 20. Each side is five, and there are four sides. $5 + 5 + 5 + 5 = 20$.

30. **A.** >. $\frac{1}{2}$ is greater than $\frac{3}{8}$.

31. **D.** 6. $9 \times 6 = 54$.

32. **A.** $237 + \underline{\hspace{1cm}} = 189$. This number sentence best describes the situation.

33. **C.** 90. $9 \times 10 = 90$.

34. **D.** 8. $8 \times 3 = 24$.

35. **A.** six buses. $54 \div 9 = 6$.

36. **D.** $\frac{3}{8}$. Three-eighths of the circle is gone.

## EXTENDED CONSTRUCTED RESPONSE QUESTION

37. In this question, there is more than one way to solve it. One way would be to line up everyone, according to the conditions of the problem. Using "b" for boy, and "g" for girl, we would line up the class like this:

    Bbb g bbb g bbb g bbb g bbb g bbb g bbb g

    This is the pattern that is described, and there are 28 students in this line-up. Thus, we find that there are 21 boys and seven girls.

    Another way of solving the problem is to observe that for each group of four students in the class, three are boys and one is a girl. That would mean that, because there are 28 students in the class, we would divide 28 by 4, and get 7. So, to get the number of boys, multiply $7 \times 3 = 21$, and for the girls, multiply $7 \times 1 = 7$.

**PART 5**

38. C. $10\frac{1}{2}$ hours. $1 + 2\frac{1}{2} + 3 + 4 = 10\frac{1}{2}$ hours.

39. D. 388. $724 - 336 = 388$.

40. A. 36. $9 \times 4 = 36$.

41. C. 52. $5 + 18 + 29 = 52$.

42. B. 92. $23 \times 4 = 92$.

43. A. 16 square centimeters. $8 \times 2 = 16$, or count the square centimeters.

44. A. 7. $31 - 24 = 7$.

45. C. 14. $5 \times e = 70 \rightarrow \frac{5}{5} \times e = \frac{70}{5} \rightarrow e = 14$.

## EXTENDED CONSTRUCTED RESPONSE QUESTION

46. One way of solving this is to divide: $324 \div 27 = 12$. Inga will use 12 gallons of gasoline to get from Boston to Newark.

    Another way of solving this problem is to subtract groups of 27 from the total until you get to zero (0):

    | | |
    |---|---|
    | $324 - 27 = 297$ | $162 - 27 = 135$ |
    | $297 - 27 = 270$ | $135 - 27 = 108$ |
    | $270 - 27 = 243$ | $108 - 27 = 81$ |
    | $243 - 27 = 216$ | $81 - 27 = 54$ |
    | $216 - 27 = 189$ | $54 - 27 = 27$ |
    | $189 - 27 = 162$ | $27 - 27 = 0$ |

    And we have 12 times, meaning that it will take 12 gallons of gas to travel from Boston to Newark.

**PART 6**

47. B. × 4. The rule is that each number is multiplied by 4.

48. A. 35. 73 − 38 = 35.

49. A. 831. 453 + 378 = 831.

50. D. 158. 250 − 92 = 158.

51. C. $\frac{1}{4}$ liter. The beaker is filled up to the $\frac{1}{4}$ liter mark.

52. D. 10. 60 ÷ 6 = 10.

53. C. 1.5 liters. 21 ÷ 14 = 1.5.

54. A. >. $\frac{3}{4}$ is greater than $\frac{5}{8}$.

## EXTENDED CONSTRUCTED RESPONSE QUESTION

55. Halstead & Company started out with 200 calculators. In the morning, they gave away 87 calculators, and in the afternoon they gave away 45. The sum of these is 87 + 45 = 132. Therefore, they gave away 132 calculators. That means that they had 200 − 132 = 68 calculators left over. Now, they signed up one person for every three calculators they gave away. That means that they signed up 132 ÷ 3 = 44 people for consultations.

**PRACTICE TEST 2**

DAY 1

Part 1

---

Directions

This part of the test contains eight short constructed response questions. You may *not* use a calculator for these questions. Write your answer on the line.

---

1.  Aiden works at a bagel store. Last week the store made 320 plain bagels, 270 sesame bagels, and 160 whole wheat bagels. How many bagels did the store make last week?

    Write your answer here ___750___

    $$\begin{array}{r} 320 \\ +\ 160 \\ \hline 480 \\ +\ 270 \\ \hline 750 \end{array}$$

2.  Consider this IN–OUT table.

    | RULE | | IN | OUT |
    |------|---|----|-----|
    | ×? | | 3 | 21 |
    | | | 6 | 42 |
    | | | 7 | 49 |

    What is the rule in this table?

    Write your answer here ___×7___

**3.** What is the area of this rectangle? The squares are centimeters.

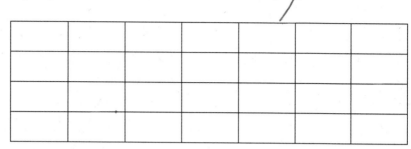

Write your answer here ___28 squre centimeters___

**4.** Matthew has a collection of Smokingwheels race cars he wants to put in carrying cases. He has 56 cars. Each case holds eight race cars. How many cases will he need to store all the race cars?

Write your answer here ___7 bags___

**5.** What is the value of $t$ in $18 + t = 29$?

Write your answer here ___t=11___

**6.** Theo's class has 19 girls and 13 boys. There are how many more girls than boys in Theo's class?

Write your answer here ___6 more girls___

**Go On**

7. What is the fact family for the multiplication/division fact triangle?

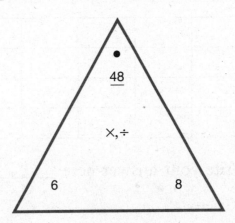

Answer:

$\underline{8} \times \underline{6} = \underline{48}$

$\underline{6} \times \underline{8} = \underline{48}$

$\underline{48} \div \underline{8} = \underline{6}$

$\underline{48} \div \underline{6} = \underline{8}$

8. Find the exact answer to 180 ÷ 3.

Write your answer here $\underline{6,360}$

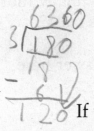

If you have time, you may review your work in this section only.

## PART 2

---

### Directions

This part of the test has 11 multiple-choice questions. You may *not* use a calculator for these questions. Fill in the circle next to your choice.

---

9. Earlier this month, 37 third graders and 28 second graders went on a field trip. How many students went on the trip in all?

    Ⓐ  55

    Ⓑ  65

    Ⓒ  9

    Ⓓ  19

    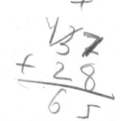

10. Mike went to Scranton's Office Supply store for some school supplies. He got five notebooks, 12 pens, 24 pencils, a pencil sharpener, an eraser, and a calendar. How many items did he get?

    Ⓐ  40 items

    Ⓑ  42 items

    Ⓒ  43 items

    Ⓓ  44 items

    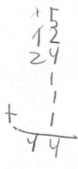

**Go On**

11. Fatima runs the ShortStop store near Hilltop high school. Last week, she sold cupcakes at a regular rate:

| Day | Monday | Tuesday | Wednesday | Thursday |
|---|---|---|---|---|
| Cupcakes sold | 13 | 26 | 39 | 52 |

If this pattern continues, how many cupcakes will Fatima sell on Friday?

Ⓐ 59

Ⓑ 63

Ⓒ 65

Ⓓ 67

12. What is the exact answer to 562 − 379?

Ⓐ 941

Ⓑ 293

Ⓒ 217

Ⓓ 183

13. What is 687 rounded to the nearest 100?

Ⓐ 700

Ⓑ 690

Ⓒ 600

Ⓓ 680

14.

In the figure above, what portion is *not* shaded?

Ⓐ $\dfrac{5}{8}$

Ⓑ $\dfrac{4}{8}$

Ⓒ $\dfrac{3}{8}$

Ⓓ $\dfrac{2}{8}$

**Go On**

15. Jennifer had a collection of hockey pucks from the games she has attended. Before the season started, she had 59 pucks. At the end of the season, she had 71 pucks. How many more pucks did she get that season?

Ⓐ 130

Ⓑ 12

Ⓒ 22

Ⓓ 120

$$\begin{array}{r} \overset{6}{\cancel{7}}\overset{11}{\cancel{1}} \\ -59 \\ \hline 12 \end{array}$$

16. Natalie went to the market to get some tomatoes. They were packaged according to the following chart:

| Tomatoes per package | Packages of tomatoes bought | Total tomatoes bought |
|---|---|---|
| 5 | 4 | ? |

Ⓐ 20

Ⓑ 16

Ⓒ 15

Ⓓ 25

17. In soccer practice, Harry took 26 shots on goal on Monday, 35 shots on goal on Tuesday, and 29 shots on goal on Wednesday. How many shots on goal did Harry take altogether in the 3 days?

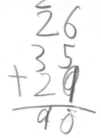

Ⓐ 70

Ⓑ 90

Ⓒ 81

Ⓓ 91

18. In the following fact triangle, what is the value of *t*?

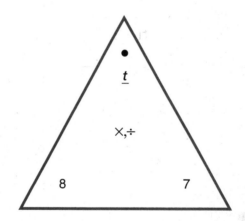

Ⓐ 49

Ⓑ 63

Ⓒ 56

Ⓓ 64

**Go On**

19.

What is the figure above?

Ⓐ A square

Ⓑ A rectangle

Ⓒ A parallelogram

Ⓓ A trapezoid

If you have time, you may review
your work in this section only.

**PART 3**

Directions

This part of the test contains eight multiple-choice questions and one extended constructed response question. You may *not* use a calculator for these questions. Fill in the circle next to your choice. Follow the directions for the extended constructed response question.

20.

What time is showing on the clock above?

Ⓐ 2:43

Ⓑ 2:48

Ⓒ 2:40

Ⓓ 2:45

**Go On**

21. Mr. Gerard has a class trip coming up; he needs to divide up the class so they can ride in cars. He has 24 students in his class, and he has six cars to put them in. How many students will go in each car?

Ⓐ 3

Ⓑ 4

Ⓒ 6

Ⓓ 8

22. Find the exact answer to 84 ÷ 7.

Ⓐ 16

Ⓑ 14

Ⓒ 12

Ⓓ 11

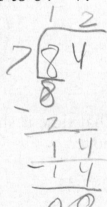

23. In the figure above, what portion is shaded?

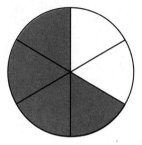

Ⓐ $\dfrac{2}{6}$

Ⓑ $\dfrac{3}{6}$

Ⓒ $\dfrac{4}{6}$

Ⓓ $\dfrac{5}{6}$

Go On

24. In the following IN–OUT table,

| RULE |
|------|
| ÷6   |

| IN | OUT |
|----|-----|
| 48 | A   |
| B  | 12  |
| 54 | C   |

what would we replace A, B, and C with?

(A) 7, 8, 9

(B) 8, 9, 10

(C) 7, 56, 8

(D) 8, 72, 9

25. Wally went with Billy to the golf course to find some golf balls. Wally found 34 and Billy found 67. How many more golf balls did Billy find?

(A) 33

(B) 37

(C) 91

(D) 101

26. Hans works in the Soda Pop Shop and can make banana splits in 4 minutes. At that rate, how long will it take him to make 11 banana splits?

Ⓐ 40 minutes

Ⓑ 42 minutes

Ⓒ 44 minutes

Ⓓ 46 minutes

27. What is the perimeter of this rectangle? Each square is a square inch.

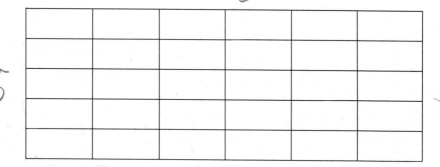

Ⓐ 36 inches

Ⓑ 22 inches

Ⓒ 30 square inches

Ⓓ 40 square inches

**Go On**

## DIRECTIONS FOR THE EXTENDED CONSTRUCTED RESPONSE QUESTION

This question is an extended response question. Remember to:

■ Read the problem and think about the answer
■ Answer all parts of the question
■ Show all your work
■ A drawing may be used to show your answer

28. At the Giant Department Store, socks come in packages of six pairs, and each package has three colors. Two pairs are gray, two pairs are blue, and two pairs are black. If Hattie wants to get eight pairs of black socks, how many packages must she buy? In total, how many pairs of socks will she have bought?

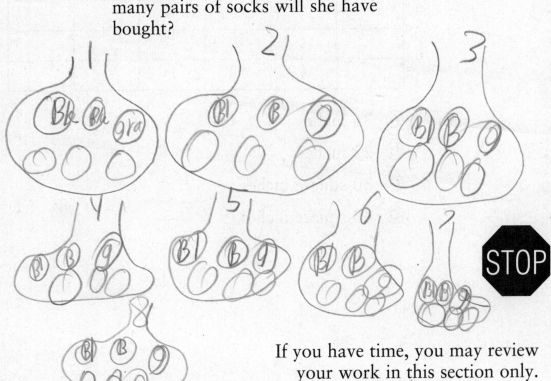

she buyed. 4 8 pairs

she buyed 8 packeges

If you have time, you may review your work in this section only.

STOP

**DAY 2**

**PART 4**

---

Directions:

This part of the test contains eight multiple-choice questions and one extended constructed response question. You *may* use a calculator for these questions. Fill in the circle next to your choice. Follow the directions for the extended constructed response question.

---

29.  Bonnie's car gets 30 miles for each gallon of gas it uses. If she takes a trip of 180 miles, how many gallons of gas will her car use?

    Ⓐ  5

    Ⓑ  6

    Ⓒ  7

    Ⓓ  8

**Go On**

**30.**

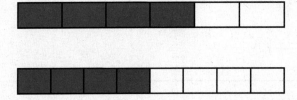

What sign should come between these two fraction bars?

Ⓐ >

Ⓑ <

Ⓒ =

Ⓓ None of the above

**31.** In the following fact triangle, what is the value of *s*?

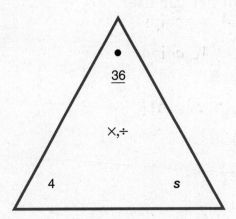

Ⓐ 9

Ⓑ 8

Ⓒ 7

Ⓓ 6

**32.** Find the exact answer to 692 − 394.

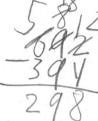

   Ⓐ 298

   Ⓑ 302

   Ⓒ 398

   Ⓓ 312

**33.** Dianna opened up her lunch and saw that she had five cookies, each having eight chocolate chips in them. How many chocolate chips were in the cookies in all?

   Ⓐ 32 chips

   Ⓑ 48 chips

   Ⓒ 40 chips

   Ⓓ 46 chips

**34.** What is the value of $d$ in: $4 \times d = 36$?

   Ⓐ 9

   Ⓑ 8

   Ⓒ 7

   Ⓓ 6

**Go On**

35. Kathy works at a bakery. She baked 64 round rolls, 82 sweet rolls, and 76 sandwich rolls. How many rolls did she bake in all?

    Ⓐ 212 rolls

    Ⓑ 112 rolls

    ● 222 rolls

    Ⓓ 232 rolls

36. Erik pitches on a baseball team. In one game last week he threw 145 pitches. In another game, he threw 59 pitches before he was replaced. How many more pitches did he throw in the first game than in the second?

    Ⓐ 88

    Ⓑ 86

    Ⓒ 84

    Ⓓ 80

## DIRECTIONS FOR THE EXTENDED CONSTRUCTED RESPONSE QUESTION

This question is an extended constructed response question. Remember to:

■ Read the problem and think about the answer
■ Answer all parts of the question
■ Show all your work
■ A drawing may be used to show your answer

37. Ms. Sullivan has kept a chart of the money she spent for the last month. Last month had 5 weeks in it, but the first week was a short one. The chart is below:

| Week | Amount spent |
|------|-------------|
| 1 | $45 |
| 2 | $60 |
| 3 | $75 |
| 4 | $70 |
| 5 | $80 |

Make a bar graph of the amount of money Ms. Sullivan spent last month. Be sure to label all the parts of the graph, and title the graph as well.

**STOP**

If you have time, you may review your work in this section only.

**Go On**

**PART 5**

### Directions

This part of the test contains eight multiple-choice questions and one extended constructed response question. You may *not* use a calculator for these questions. Fill in the circle next to your choice. Follow the directions for the extended constructed response question.

38.  Laura babysits the children next door. In January, she babysat for a total of 56 hours. In February, she babysat a total of 37 hours. How many more hours did she babysit in February than in January?

Ⓐ  93 hours

Ⓑ  13 hours

Ⓒ  21 hours

Ⓓ  19 hours

can not

do

39. Lydia bought some cookies for her class. Cookies are packaged according to the following table:

| Cookies per package | Packages bought | Total cookies bought |
|---|---|---|
| 24 | ? | 96 |

Using the table, calculate how many packages Lydia bought.

Ⓐ 4

Ⓑ 6

Ⓒ 5

Ⓓ 7

40. Consider the following IN-OUT table:

| RULE |
|---|
| +? |

| IN | OUT |
|---|---|
| 5 | 12 |
| 13 | 20 |
| 27 | 34 |

In this table, what is the rule?

Ⓐ +5

Ⓑ +7

Ⓒ +9

Ⓓ +6

**Go On**

**41.** What is the perimeter of this rectangle? The squares are feet.

4

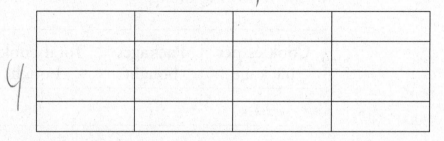

4

Ⓐ　4 feet

Ⓑ　14 feet

Ⓒ　12 feet

Ⓓ　16 feet

**42.** Mike and Maya went out to the golf range to practice their driving. Each bucket they got had 30 golf balls in it. If they got 13 buckets between them and they hit them all at the range, how many balls did they hit?

Ⓐ　390

Ⓑ　380

Ⓒ　370

Ⓓ　360

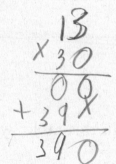

43. What is the value of $k$ in $k - 55 = 93$?

    Ⓐ  42

    Ⓑ  38

    Ⓒ  148

    Ⓓ  138

44. Samantha and Ginger sold girl scout cookies in front of their local supermarket. Samantha sold 350 cookies and Ginger sold 580 cookies. How many cookies did they sell altogether?

    Ⓐ  130 cookies

    Ⓑ  920 cookies

    Ⓒ  230 cookies

    Ⓓ  930 cookies

**Go On**

45. In the following fact triangle, what is the value of *p*?

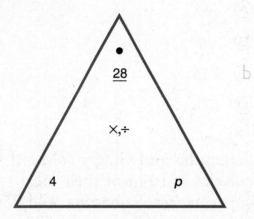

(A) 9

(B) 8

(C) 7

(D) 6

## DIRECTIONS FOR THE EXTENDED CONSTRUCTED RESPONSE QUESTION

This question is an extended constructed response question. Remember to:

■ Read the problem and think about the answer
■ Answer all parts of the question
■ Show all your work
■ A drawing may be used to show your answer

46. David was walking along the seashore collecting seashells. He collected 30 shells. He then wanted to arrange them in a rectangular pattern. Show at least three ways David could display his shells.

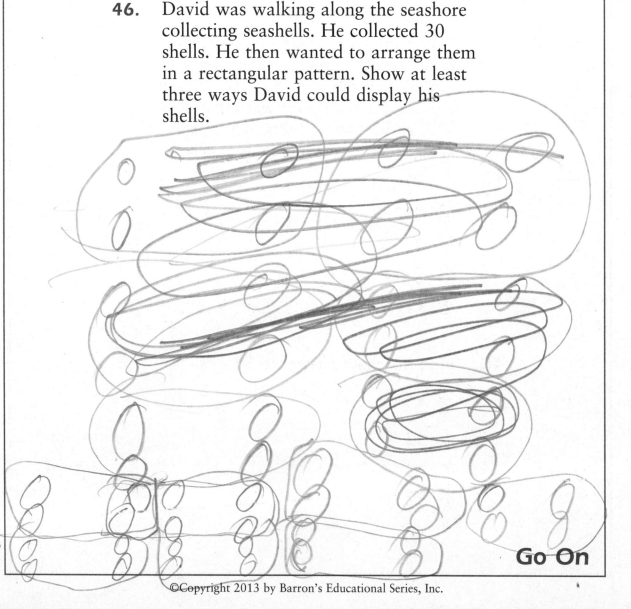

**Go On**

## PART 6

### Directions

This part of the test contains eight multiple-choice questions and one extended constructed response question. You may *not* use a calculator for these questions. Fill in the circle next to your choice. Follow the directions for the extended constructed response question.

47. Bebe stopped at a garage sale and found some things she liked. She found a bookshelf that cost $55, a sofa that cost $70, a set of glassware that cost $35, and a toaster oven that cost $15. How much did Bebe pay for these items?

Ⓐ $175

Ⓑ $170

Ⓒ $185

Ⓓ $180

48. Find the exact answer to $91 \div 7$.

Ⓐ 17

Ⓑ 12

Ⓒ 13

Ⓓ 15

49. The Hill St. School had 428 students in it. The Arthur School had 279 students. How many more students were in the Hill St. School?

Ⓐ 259 students

Ⓑ 251 students

Ⓒ 135 students

Ⓓ 149 students

50. Find the exact answer to $8 \times 60$.

Ⓐ 420

Ⓑ 480

Ⓒ 400

Ⓓ 440

51. What symbol should go between these two fractions?

$$\frac{3}{4} \underline{\hspace{2cm}} \frac{3}{8}$$

Ⓐ >

Ⓑ <

Ⓒ =

Ⓓ None of the above

**Go On**

**52.** In the following fact triangle, what is the value of *h*?

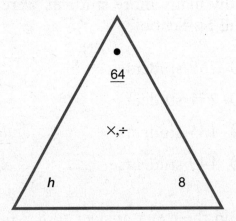

Ⓐ 5

Ⓑ 6

Ⓒ 7

Ⓓ 8

**53.** Jane planted some flowers for her mother, who loves pansies. She planted 23 pansy plants in the front yard, 19 pansy plants in the side yard, and 32 pansy plants in the backyard. How many pansy plants did she plant altogether?

Ⓐ 86 pansies

Ⓑ 82 pansies

Ⓒ 74 pansies

Ⓓ 70 pansies

54. Vince went to Barney's Bargain Basement to get some hot dog rolls. They are packaged 12 to a bag according to the following table:

| Rolls per package | Packages bought | Total rolls bought |
|---|---|---|
| 12 | ? | 120 |

How many packages did Vince buy?

Ⓐ   8 packages

Ⓑ  10 packages

Ⓒ  12 packages

Ⓓ  14 packages

**Go On**

## DIRECTIONS FOR THE EXTENDED CONSTRUCTED RESPONSE QUESTION

This question is an extended constructed response question. Remember to:

- Read the problem and think about the answer
- Answer all parts of the question
- Show all your work
- A drawing may be used to show your answer

55. Jim needs to buy some little bottles of Erase-It correction fluid. He wants to buy 15 bottles. Jim saw that the bottles were selling for three for $4.00 or five for $6.00. Which is the better buy? Show two ways of finding this.

1.  $6.00 ⎫                    $4.00
    $6.00 ⎬ 15                 $4.00
    $6.00 ⎭ bottles            $4.00
                               $4.00
    $18.00        +           $4.00

         less                 $20.00

2. This says five five
   is more than three.

STOP

If you have time, you may review your work in this section only.

# ANSWERS TO PRACTICE TEST 2

## DAY 1

## PART 1

1. The bagel store made $320 + 270 + 160 = 750$ bagels.

2. All three numbers are multiplied by 7. The rule is $\times 7$.

3. There are 28 square centimeters. $7 \times 4 = 28$, or count the squares.

4. $56 \div 8 = 7$. Matthew will need seven cases to store all his race cars.

5. $18 + t = 29 \rightarrow 18 - 18 + t = 29 - 18 \rightarrow t = 11$.

6. $19 - 13 = 6$. There are six more girls than boys in Theo's class.

7. The fact family for this fact triangle is:

   a.  $6 \times 8 = 48$

   b.  $8 \times 6 = 48$

   c.  $48 \div 6 = 8$

   d.  $48 \div 8 = 6$

8. $180 \div 3 = 60$.

## PART 2

9. B. 65. $37 + 28 = 65$.

10. D. 44 items. $5 + 12 + 24 + 1 + 1 + 1 = 44$.

11. C. 65. Fatima sells 13 more each day, and $52 + 13 = 65$.

12. D. 183. $562 - 379 = 183$.

13. A. 700. 687 rounded to the nearest 100 is 700.

14. C. $\frac{3}{8}$. $\frac{5}{8}$ is shaded, so $\frac{3}{8}$ is not shaded.

15. B. 12. 71 − 59 = 12.

16. A. 20. 5 × 4 = 20. Natalie bought 20 tomatoes.

17. B. 90. 26 + 35 + 29 = 90.

18. C. 56. 8 × 7 = 56.

19. D. A trapezoid. A four-sided figure with two sides parallel to each other is a trapezoid.

**PART 3**

20. A. 2:43. The time on the clock is 2:43.

21. B. 4. 24 ÷ 6 = 4. Mr. Gerard will put four students in each car.

22. C. 12. 84 ÷ 7 = 12.

23. C. $\frac{4}{6}$. $\frac{4}{6}$ of the circle is shaded.

24. D. 8, 72, 9. In the IN–OUT table, 48 ÷ 6 = 8; 72 ÷ 6 = 12; 54 ÷ 6 = 9.

25. A. 33. Billy found 67 − 34 = 33 more golf balls than Wally did.

26. C. 44 minutes. Hans can make 11 banana splits in 44 minutes.

27. B. 22 inches. The perimeter is 5 + 6 + 5 + 6 = 22. Or, it is 2 × (5 + 6) = 22.

## EXTENDED CONSTRUCTED RESPONSE QUESTION

**28.** For this question, we need to find the number of black socks and the total number of socks bought.

  **a.** One way of solving this is to consider that, in each package of six pairs of socks, there are two pairs of black socks. If Hattie wants eight pairs of black socks, she will need to buy: $2 \times 4 = 8 \rightarrow$ four packages of socks. If Hattie buys four packages of socks, then she will have: $4 \times 6 = 24 \rightarrow$ 24 pairs of socks.

  **b.** Another way to solve this is to draw a picture.

As can be seen, for Hattie to get eight pairs of black socks, she will need to get four packages, each having two pairs of black. That's 24 pairs of socks.

## PART 4

**29.** B. 6. $180 \div 30 = 6$. Bonnie will need six gallons of gas to go 180 miles.

**30.** A. $>$ $\dfrac{4}{6}$ is greater than $\dfrac{4}{8}$.

**31.** D. 9. $4 \times 9 = 36$.

**32.** A. 298. $692 - 394 = 298$.

**33.** C. 40 chips. $5 \times 8 = 40$.

**34.** A. 9. $4 \times d = 36 \rightarrow \dfrac{4}{4} \times d = \dfrac{36}{4} \rightarrow d = 9$.

**35.** C. 222. $64 + 82 + 76 = 222$ rolls.

**36.** B. 86. $145 - 59 = 86$.

## EXTENDED CONSTRUCTED RESPONSE QUESTION

**37.**

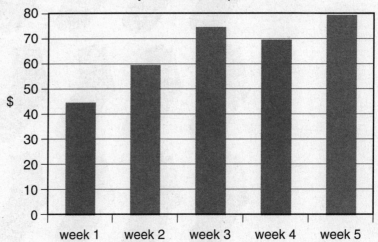

Money Ms. Sullivan spent last month

**PART 5**

**38.** D. 19 hours. $56 - 37 = 19$.

**39.** A. 4. $24 \times 4 = 96$.

**40.** B. + 7. In all of the cases, 7 is added to the IN value.

**41.** D. 16 feet. $4 + 4 + 4 + 4 = 16$. Or, $4 \times 4 = 16$.

**42.** A. 390. $30 \times 13 = 390$.

**43.** C. 148. $k - 55 = 93 \rightarrow k - 55 + 55 = 93 + 55 \rightarrow k = 148$.

**44.** D. 930 cookies. $350 + 580 = 930$.

**45.** C. 7. $4 \times 7 = 28$.

## EXTENDED CONSTRUCTED RESPONSE QUESTION

46. There are many ways of arranging the seashells. Here are four ways:

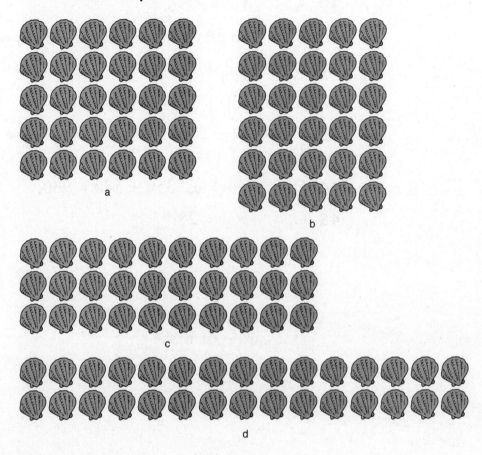

a. 5 rows of 6 shells each.

b. 6 rows of 5 shells each.

c. 3 rows of 10 shells each.

d. 2 rows of 15 shells each.

Any of these arrangements are good ways for David to arrange the 30 seashells he collected.

**PART 6**

**47. A.** $175. $55 + $70 + $35 + $15 = $175.

**48. C.** 13. 91 ÷ 7 = 13.

**49. D.** 149 students. 428 − 279 = 149.

**50. B.** 480. 8 × 60 = 480.

**51. A.** >. $\frac{3}{4}$ is greater than $\frac{3}{8}$.

**52. D.** 8. 8 × 8 = 64.

**53. C.** 74 pansies. 23 + 19 + 32 = 74.

**54. B.** 10 packages. 12 × 10 = 120.

## EXTENDED CONSTRUCTED RESPONSE QUESTION

**55.** To solve this problem, we could calculate the price of one bottle of Erase-It:

  **a.** $4.00 ÷ 3 = 1.333. Each bottle will be $1.33, at three for $4.00.

  $6.00 ÷ 5 = 1.2. Each bottle will be $1.20, at five for $6.00.

  **b.** A second way to solve this would be to calculate the price of 15 bottles:

  At three for $4.00, 15 bottles would be $20.00 (because we multiplied 3 by 5, we must multiply $4.00 by 5).

  At five for $6.00, 15 bottles would be $18.00 (because we multiplied 5 by 3, we must multiply $6.00 by 3).

# INDEX

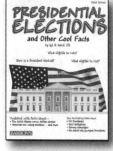